BOTANY FOR
GARDENERS

BOTANY FOR GARDENERS

An Introduction and Guide
by
Brian Capon

TIMBER PRESS
Portland, Oregon

© 1990 by Timber Press, Inc.
All rights reserved
Fifth casebound printing 1996
Fifth paperback printing 1996

ISBN 0-88192-163-7 (cased)
ISBN 0-88192-258-7 (paper)
Printed in Hong Kong

TIMBER PRESS, INC.
The Haseltine Building
133 S.W. Second Avenue, Suite 450
Portland, Oregon 97204-3527, U.S.A.

Library of Congress Cataloging-in-Publication Data
Capon, Brian.
    Botany for gardeners : an introduction and guide / by Brian Capon.
        p. cm.
    Includes bibliographical references.
    ISBN 0-88192-163-7 (hardback)
    ISBN 0-88192-258-7 (paperback)
    1. Botany.   2. Gardening.   I. Title.
QK50.C36   1990
581—dc20
                                                        89-20340
                                                          CIP

Contents

TO MY STUDENTS

Acknowledgments

Over the years, I have had the pleasure of introducing countless numbers of students to the science of botany. Recently, the desire to relate the wonders of plant life to an even broader audience led to thoughts of a botany book for gardeners.

The concept was nurtured by suggestions from my good friends and plant enthusiasts: Dr. Willard Van Asdall of the University of Arizona, Eva Morgan, Dr. Larry Palkovic, and Diane Palkovic. For the transformation of a basic outline of ideas into the present volume, I am indebted to Richard Abel of Timber Press, whose enthusiasm for the project and valuable editorial advice made the manuscript's preparation a thoroughly enjoyable experience.

I wish to thank Dr. Sam Poolsawat for the electron micrograph preparation in Chapter 8.

Finally, thank you to friends and neighbors who were supportive of the work and gladly provided photographic subjects from their gardens.

Brian Capon
Los Angeles, California.
February, 1989.

Introduction

Earth has been called the Green Planet; in the vast reaches of the solar system, perhaps the universe, it is a solitary world uniquely clothed in a mantle of vegetation. And because of its plants, other forms of life are able to inhabit this place. From simple beginnings, plants evolved first among earth's living things, thus setting a priority that still abides: Plants, in one form or another, can exist forever without animals, but animals cannot exist without plants.

Plants purify the air; they exchange the oxygen that we breathe with carbon dioxide, a poisonous gas in too high a concentration. Plants convert the energy of sunlight into foods that sustain all animals and, from the soil, draw minerals—nitrogen, potassium, calcium, iron—that are essential for our growth and continued health. For creatures large and small, plants provide shade from the sun, refuge from predators and protection from the most destructive aspects of earth's climate.

Since the first cells came into being millions of years ago, plants have been the connecting links in an unbroken chain of life. It is they that have made the biosphere, the part of earth's crust where both plants and animals exist, a vibrant and constantly changing place offering limitless opportunities for the inquiring mind to explore.

The range of uses we make of plants is as broad as our ingenuity permits. We have exploited them for fibers to make cloth, drugs to cure a multitude of ailments, wood to construct houses, furniture and ships. From them we have extracted raw materials to manufacture innumerable goods, including paper.

Without that latter commodity, the detailed history of our race would not have been recorded and so remembered, nor could knowledge have been so easily disseminated. And culture, the possession of which makes humans out of animals, would never have developed beyond the basic skills and habits of primitive peoples had we not had paper on which to write music, poetry and prose.

Some of us look at plants as a source of livelihood, while others find them intriguing subjects for scientific study. But most enjoy plants for the sheer delight of having them in their every-day surroundings, to savor the varied colors, textures, tastes and aromas that they alone can offer. Plants stimulate the senses, give peace to the weary mind, and satisfy man's spiritual being in search for answers to the mysteries of life.

Few gardeners share the botanist's knowledge of plant biochemistry, anatomy, physiology and intricate reproductive systems, yet all have experienced the extraordinary satisfaction derived from growing flowers, fruits, vegetables and trees.

When we work with plants, questions about them inevitably come to mind. What takes place inside a seed after we have set it in the ground? How does water travel from soil to treetops? What makes a plant become bushy with repeated pruning? What controls seasonal flowering patterns? How do plants grow, and why is light necessary to make growth happen? Over the centuries, botanists have worked to find answers to these and other problems. Slowly, plants have revealed some of their secrets.

Botany is a useful and rewarding study from which, unfortunately, many laypersons are frightened away by the technical jargon that constitutes the "language" of the science. The reader will encounter a number of scientific words in the following pages. Some are part of the common parlance of gardeners. For want of suitable non-technical equivalents, others cannot be avoided when writing such a book.

Each technical word, whether common or obscure, is explained in the text and Glossary, and occasional reference is made to the Greek and Latin roots from which these words have been derived. In addition, it is hoped that the numerous illustrations will give added meaning to the botanical vocabulary and ideas developed. The Glossary may be used as both a dictionary of botanical words and an index for the body of the work, since page references have been added to the definitions.

Some of the photographic subjects are not the customary things that gardeners look for in plants. But they are plants or parts of them seen in close-up, sometimes through a microscope. A majority of the plant specimens that have been photographed were selected from those available in my own and neighbors' gardens, local parks and botanical gardens in Southern California. But the broad principles of botany each exemplifies are equally applicable to plants in almost any part of the world.

There are close to 400,000 recognizably different kinds of plants, called *species,* in the world today. So diverse are their forms that to write an all-inclusive definition of the word "plant" is not at all easy. One-third of all plants do not have roots, stems and leaves as we know these parts in the examples most familiar to us. About 150,000 plant species never produce flowers, and almost that same number do not grow from seeds, but rather from dustlike particles called spores. Although the vast majority of plants manufacture their own food supplies by a process called photosynthesis, mushrooms, molds and other fungi—which some biologists include in the Plant Kingdom— rely on foods created by green plants for their sustenance, as do animals. Most plants spend a lifetime anchored in one place, yet a few simple, one-celled forms are capable of swimming to different locations in the waters they inhabit.

It is this kind of diversity and amazing variety of shapes, colors and lifestyles that continually excite our interest in these organisms called plants.

As we delve into the science of botany, we shall largely be concerned with the two groups of plants with which we, as gardeners, most often work. One, known as the flowering plants or Angiosperms, is the largest group in the Plant Kingdom and consists of about 250,000 species. The name "Angiosperm" refers to the fact that seeds from these plants are formed inside containers that we call fruits (Greek: *angeion,* vessel; *sperma,* seed). The flowering plants most often decorate our homes and landscapes, supply almost all of the vegetable matter in our diets, and are the source of the world's hardwoods. They are the most sophisticated of plant forms and are best adapted to survive in a wide range of climates and places.

Second are Gymnosperms, plants that produce seeds in the open spaces of cones—between the flaplike parts that make up a pine cone, for example. The Greek words *gymnos,* "naked" and *sperma,* "seed" describe this form of development. On the evolutionary scale, Gymnosperms are more primitive than Angiosperms but are of considerable economic importance as well as interest to landscapers for their compact forms and richly-colored, needle-shaped or scalelike leaves. Softwoods such as pine and fir are not only used to make paper, lumber, plywood, etc., but are the source of a group of products called naval stores—pitch, turpentine and rosin. The Gymnosperms include all the conifers: cedar, redwood, juniper, cypress, fir and pine, and the largest living things on earth, the giant sequoias. Many ornamental shrubs, including varieties of *Chamaecyparis* (False Cypress) and *Thuja occidentalis* (American Arborvitae) are members of this group and, least typical of Gymnosperms, Cycads and the beautiful Maidenhair tree, *Ginkgo biloba,* a broadleaved species.

For comparative purposes, passing mention is made of ferns, mosses, mushrooms and other primitive plants, but it is to the flowering plants and Gymnosperms that we direct our attention since it is they that give us the most revealing picture of how marvelous plants are.

I. GROWTH

Prologue

A person, plant, pebble, this book page—four objects no one has trouble putting into simple categories of living and non-living. But what makes the difference? Why can we be so sure that the potted geranium is living and a piece of the same plant that was pressed and dried last year is unquestionably dead? From all appearances, a seed also seems dead. What happens when, upon reawakening, it becomes charged with life during germination, and what mysterious entity leaves a plant when it dies? Few people believe in the departed souls of plants. Philosophers may endlessly ponder such questions, but trying to find answers strictly based on observation and repeatable experiment are the essence of scientific inquiry.

To begin, a living plant has the ability to make flowers, and seeds from which other plants of the same species can be grown. In other words, a living plant can reproduce. A dead one has lost that capacity. Then again, if one has the opportunity to look at any part of a plant through a microscope, it becomes obvious that plants are composed of countless numbers of cells, invisible to the naked eye. This gives another clue to the nature of living things. It may be argued that cells are still in a leaf when it is dead and dried. But when the leaf was a part of a living plant, its cells were actively engaged in a complicated chain of chemical reactions, grouped together under the term "metabolism". We can be quite sure that, as long as a cell or a whole creature is alive, it is going to display some sort of metabolic activity. When their chemistry irreversibly stops, cells die.

Perhaps the most obvious difference between a rock and a rose is that the rock doesn't grow. In fact, it progressively becomes smaller as its surface erodes. Plants and animals, on the other hand, begin life as single, fertilized eggs and become larger as they mature. In the case of animals, including ourselves, a *determinate growth* pattern dictates a pre-fixed, maximum size that the body may reach. This pattern is implicit in and established by *genes*, cellular instructions inherited from parents, and is more or less related to the number of cells that the body is programmed to produce. Strenuous exercise may enlarge cells but relatively few new cells are added. Full growth potential is realized if an animal receives adequate nutrition and its muscles are exercised, especially during the formative years.

For the most part, plants have no definite size toward which they grow. That is, they display *indeterminate growth* or, at least, their stems and roots do. When left untouched and growing in an unrestricted volume of soil, a plant's roots will never reach an "established size," nor will its branches in the freedom of an open-air space. Limits of plant growth are proportional to the availability of light, water, minerals and oxygen. Lifespan is genetically determined—one year for annuals, two for biennials and indefinitely in perennials.

Compare the indeterminate growth pattern of roots and stems with that found in leaves, flowers, fruits and seeds. The latter are both characteristically ephemeral and determinate in growth. Their maximum possible sizes are rarely displayed in nature but can be realized under a skillful gardener's control of the plant's environment. With plenty of fertilizer, careful watering schedules, optimum illumination and thinning—removal of parts that may compete for available nutrients—a plant can be pushed to the limits of leaf, flower and fruit growth. All regular visitors to county fairs know what prize-winning blossoms and fruits look like. They are giants compared with normal specimens but never reach super-giant status.

Animals grow up and spend their lives in a variety of places. Mobility enables them to choose habitats that are most favorable for existence under changing conditions at different times of the year. A plant is anchored in one place throughout its life. Half of its body, its *root system,* is buried in the dark, damp and somewhat stuffy recesses of the soil. Despite being surrounded by a legion of potentially destructive grubs and soil microorganisms—fungi and bacteria—from which the roots can't escape, and their passage through the soil manipulated by encounters with immovable rocks, roots are wonderfully adapted to this strange, hidden environment.

In contrast, *shoot systems,* consisting of stems and leaves, occupy a sunlit, airy but frequently tempestuous world. Growth impediments are different from those below ground and may range from insects and larger animals with voracious appetites, out to survive at the plant's expense, to the drying effect of wind and sun, or even damage from fire.

Roots and shoots are frequently thought of as different entities growing in opposite directions. To a plant, they are parts of the whole body that must be as well coordinated as are torso and legs during the course of growth and the varied activities humans undertake. Root growth and shoot growth are harmonized events, one complementing the other, with energy reserves and raw materials for body building equally allocated to the two halves. And when daily or seasonal environmental changes affect one part, the other must respond in sympathy. Fulfillment of the fundamental qualities of living things—*reproduction, cellular metabolism* and *growth*—can only be achieved by such precisely controlled interactions between roots, stems, leaves and flowers.

This section deals with plant cells, their structure and role in growth processes, followed by a look at the growth of representative flowering plants from germinated seeds to maturity. For convenience, growth and the external forms of roots, stems and leaves are treated under separate headings. This is not to imply that all parts of a plant do not develop simultaneously.

The science of botany is divided into various disciplines, each having its specialists, subject limitations and technical vocabulary. Among them, *cytology* (Greek: *kytos,* container) is the detailed study of cells. Study of the form and structure of plants is the work of morphologists (Greek: *morphe,* form). By virtue of their practical relationships with plants, gardeners are more familiar with *morphology* than with cytology. The reader is invited into the realm of cells to better understand what goes on inside roots, stems and leaves when they grow.

Chapter 1

Cells and Seeds.
Basics and Beginnings

Cells

Robert Hooke, an English physicist, was understandably excited when, in 1665, he wrote about having used a crude microscope to look at a slice of cork. He probably thought he'd simply confirm the prevailing idea that plants are composed of some sort of amorphous material, like clay shaped by the Creator's hands. But contrary to such expectations, Hooke was the first person to find that plants were actually constructed of tiny units which he named "cells". His choice of word more likely reflected his acquaintance with Latin (*cella,* a small room) than with the interior of a jailhouse.

What subsequently became known as the Cell Theory—that all living things are composed of one or more cells—was as revolutionary to scientific thought as was, in our own time, the discovery of DNA (deoxyribonucleic acid), the chemical substance controlling biological inheritance. Each year, as scientists delve deeper into cells, the revelation of what life really is, at microscopic and finer levels, continues to offer surprises.

To get an idea of what a typical plant cell is like and what it can do, think of a large factory, capable of manufacturing thousands of different and elaborate products from simple raw materials—water, air and soil. The factory uses sunlight rather than electricity or oil as an energy source. It is designed to exert considerable autonomous control over what goes on within its boundaries and, whenever increased productivity is called for, it simply builds an exact copy of its entire physical structure—within a day or two. Now, mentally squeeze the factory into a box, each side approximately 1/2,000 of an inch (0.05 mm). That is a cell.

The living part of a cell, the *protoplasm,* consists of two parts: a *nucleus,* which is the center of inheritance and cellular control, and the *cytoplasm,* a soft, jelly-like material (a colloid) in which most of the cell's metabolism takes place. The cytoplasm is enclosed within a sac called the *cytoplasmic membrane.* This, like other membranes in a cell, is

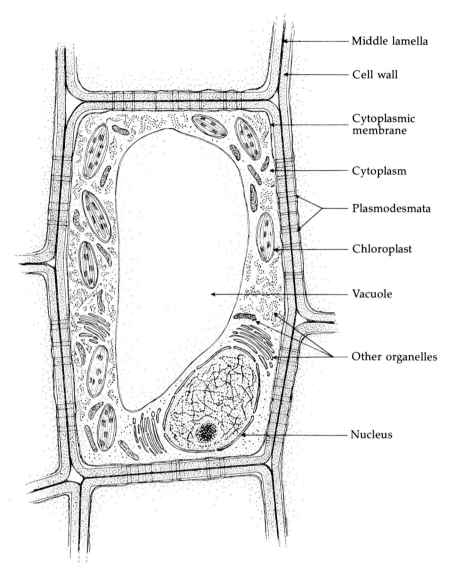

Middle lamella

Cell wall

Cytoplasmic membrane

Cytoplasm

Plasmodesmata

Chloroplast

Vacuole

Other organelles

Nucleus

Details of a plant cell.

composed of protein and fatty substances and has the ability to control the passage of water, foods and selected minerals across the boundary that it defines.

Suspended in the semi-liquid cytoplasm are numerous small bodies, or *organelles*, which specialize in the cell's separate functions. Some organelles are the same in both plant and animal cells, hinting at ancient ancestral ties. *Chloroplasts* are organelles unique to plants since they are the place where photosynthesis takes place, where light energy is used to manufacture foods. The green pigment *chlorophyll,* essential for the process, is located within the chloroplasts, as its name indicates (Greek: *chloros,* green; *plastos,* body; *-phyll,* leaf). Obviously, one would not expect to find chloroplasts in most roots or other parts of a plant that are not green. The color of a leaf is actually the combined appearance of millions of chloroplasts discernible only with the aid of a microscope.

Other organelles include *mitochondria* that extract energy from foods by a process called *cellular respiration* and those that specialize in protein production, the *ribosomes.* The functions of some organelles, visible only with powerful electron microscopes, are still not fully understood.

The nucleus of a cell is its control center from which "instructions" for the cell's operation, maintenance and reproduction emanate. It is comparable to the main office in the imaginary industrial plant. Inherited *genes,* composed of the DNA mentioned earlier, are located in the nucleus. These are the blueprints for making more cellular factories.

A *vacuole* occupies a large part of the volume of most plant cells. Although the word vacuole means "empty space", it is a membrane-bound inner sac containing much of a cell's stored water and serves as a repository for excess mineral nutrients as well as toxic waste products from the cell's metabolism.

Each cell is designed to function most of the time as an independent unit. Yet their metabolism and other activities are enhanced when groups of cells act in concert by the exchange of foods and other materials by way of inter-connecting stands of cytoplasm, called *plasmodesmata* (Greek: *desmos,* chain).

Cell Walls

The protoplasm of each cell is surrounded by a rigid *cell wall* that protects the living contents. Between adjacent cell walls, the substance *pectin* forms a thin layer, called the *middle lamella* (a sheet), which binds the cells together. This same substance, when commercially extracted from plants and sold in supermarkets, is used to thicken jams and fruit jellies.

Collectively, cell walls give structural support to a plant, the degree of rigidity of any part being proportional to the relative thickness of its constituent cells' walls. The light-weight, delicate structure of a leaf, for example, indicates that it is composed of thin-walled cells; whereas in woody stems supporting heavy loads, cells with extra-thick walls are developed.

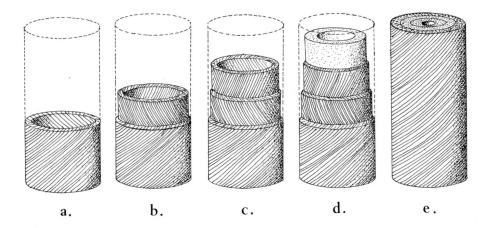

Thickening of a plant cell wall. a. Cellulose microfibrils of the primary cell wall. The cavity at the center is occupied by the protoplasm (not shown). b. and c. Secondary wall thickening begins with additional layers of cellulose; each layer is laid down in a different direction. d. A layer of lignin is added. e. With the addition of another lignin layer, wall thickening is complete. The shrunken protoplasm soon dies.

When a cell is first formed, its wall is thin and largely composed of the substance *cellulose.* This is the cell's so-called *primary wall.* With increased age, the wall may thicken by addition of more cellulose and, ultimately, by the introduction of a hardening substance called *lignin.* Hardwoods like oak and ash are made up of cells with heavily lignified walls. All of these extra layers constitute the cell's *secondary wall.*

Cellulose is laid down in microscopic threads called *microfibrils;* lignin forms deposits on the cellulose surface. Each new layer of wall material, produced by the living cytoplasm, is set in place inside the previously-formed layer.

Obviously, as walls thicken, the space occupied by the living contents decreases and the ability of water and oxygen to reach the cytoplasm is diminished. It is literally an act of suicide that kills the protoplasm and ends wall thickening. Even so, the remaining hollow cell walls continue their supportive roles throughout the life of the plant. Most people are surprised to learn that, in a living tree, as much as 98% of its trunk and branches are composed of dead cells, including those that conduct water.

Wall Structure and Cell Growth

Most cells in a plant, especially those in roots and stems, grow in a specific direction, dictated by the way in which cellulose microfibrils are arranged in the walls. If one thinks of a cell as a slightly elongated box, placed in an upright position, the four sides are formed from microfibrils placed parallel to one another and coiled around the box. Microfibrils in the top and bottom have a very different, criss-cross pattern.

When a cell enlarges, its walls temporarily soften. At the same time, cytoplasmic swelling takes place as a result of water uptake. Bonds between side-wall microfibrils are loosened and the cellulose threads are spread apart by the internal pressures. Since the microfibrils in end walls are interwoven, similar stretching is not possible. That is why cells principally grow in length, paralleling the general direction of vertical growth of stems and roots. Thickening of these plant parts results from a different growth process that shall be discussed later.

Once a cell reaches a pre-determined maximum length, the addition of secondary wall thickenings prevents further enlargements.

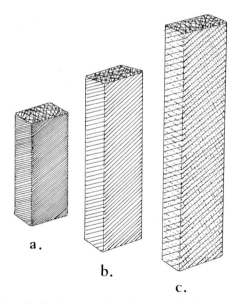

a.

b.

c.

How cellulose microfibrils determine the direction of cell growth. a. The side walls of a young cell have microfibrils arranged in parallel formation. b. The cell elongates when microfibrils in the side walls spread apart from internal pressures within the cell; the criss-cross pattern of end-wall microfibrils prevents the cell from growing in width. c. Having reached its maximum length, secondary wall thickening prevents further growth in the cell's length.

Growth Processes

Two processes taking place at a cellular level contribute to a plant's growth. In the first, new cells are formed by the division of cells already in the plant body. Each time a cell divides, two complete cells are produced. Every cell in a plant, with the exception of the original fertilized egg, has had its origin in this process.

The most important part of cell division is providing each new cell with a nucleus containing a complete set of genes. This is accomplished during a process called *mitosis* (Greek: *mitos,* thread) in which the nuclear DNA becomes organized into sets of thread-like structures called *chromosomes.* (Literally, the word chromosome means "colored body", from the fact that they readily stain with artificial dyes.) The chromosomes go through an elaborate "dance" sequence, culminating in matched chromosome parts being segregated into the two newly developed cells.

The second growth process in plants was described in the previous section where cells undergo a limited period of elongation.

Meristems

The two phases of growth, described above, occur in well defined places within a plant rather than as scattered, random events. Cells divide in areas called *meristems* (Greek: *meristos,* divided); close-by lie regions of cell enlargement.

At the tip (apex) of each stem and root an *apical meristem* contributes cells to the length of these plant organs. Such increases in stem and root length, before thickening, are referred to as the plant's *primary growth* process. What primary growth does is to ensure that leaves are quickly elevated into sunlight and roots penetrate deeply into the soil. The rapid growth of a seedling, after it has emerged from the soil, is a familiar display of primary growth that continues as long as roots and stems lengthen.

When stems have gained moderate height, it is important that they begin to thicken toward their bases to give added stability and support for the leaf mass. This is called *secondary growth* and results from cell divisions in meristems located inside, throughout the length of the stems. These *lateral meristems* also extend into the roots of larger plants. Secondary growth in a tree creates the slow but measurable thickening of its trunk and branches as well as the upper portions of roots that may emerge above the soil surface.

During seasons of active growth, both the apical and lateral meristems make their separate but coordinated contributions to the shape and size of a plant. When one looks at trees from day to day,

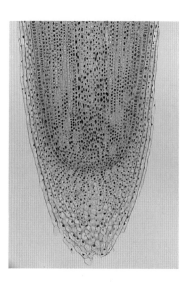

Mitosis—the dance of the chromosomes. The root's apical meristem is the U-shaped area a short distance from the tip (photograph at left). In the meristem's cells, division of the nuclei involves the separation of matching chromosome strands in stages named: a. prophase, b. metaphase, c. anaphase, and d. telophase. e. Two daughter cells. Several of these stages are shown in the photograph of the meristem's cells, greatly magnified.

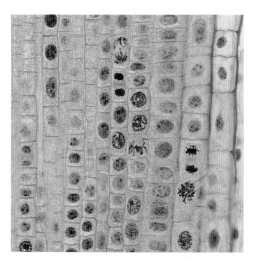

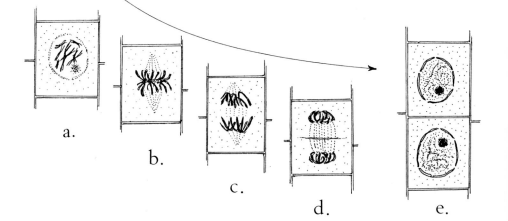

a.

b.

c.

d.

e.

changes that have taken place are hardly noticeable. Only at pruning time does the reality of plant growth strike home. More impressive is the experience of returning to old haunts after several years of absence and hardly recognizing once familiar trees, now transformed by time from saplings into stately patriarchs.

Cell division and cell enlargement, the basics of plant growth, are simple in principle yet complex in their undertaking. Finding what happens during growth in terms of the cell's chemistry, beyond the limits of human sight, is the principal preoccupation of today's scientists. Having explained such matters, will we be closer to understanding the secrets of life?

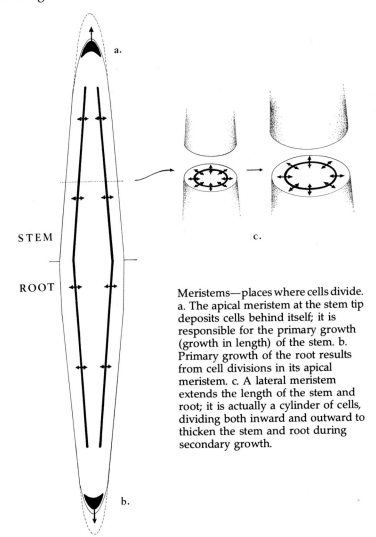

STEM

ROOT

c.

a.

b.

Meristems—places where cells divide. a. The apical meristem at the stem tip deposits cells behind itself; it is responsible for the primary growth (growth in length) of the stem. b. Primary growth of the root results from cell divisions in its apical meristem. c. A lateral meristem extends the length of the stem and root; it is actually a cylinder of cells, dividing both inward and outward to thicken the stem and root during secondary growth.

Seed Coats

No animal can lay claim to being dead before it became alive. Yet it is one of Nature's contradictions that flowering plants and Gymnosperms should begin their lives as seeds, the cells of which show no sign of metabolic activity and, therefore, are technically dead. We get around this dilemma by saying that, correctly, a *viable* seed—one ultimately capable of germination—is only *dormant,* not truly dead.

Given the right conditions, a seed wakens from its dormancy and enters a short period of intense activity, the likes of which are not to be repeated at any other time during the plant's life history. Watching the entire process of seed germination unfold within a matter of days, is one of the awe-inspiring events to which only plants-people are privy.

Seeds are extraordinary objects. They are compact, easily stored and capable of survival through freezing temperatures or prolonged drought, conditions that usually kill the parent plants. When kept dry, seeds resist fungal attack and, although they contain rich stores of food attractive to animals, they frequently elude predators by being a drab, brown color that offers camouflage against the background of the soil.

The "skin" of a seed is its *seed coat,* the color, texture and thickness of which vary from species to species. Thickness and hardness of the seed coat determines how fast water can penetrate it. This, in turn, relates to how soon germination may take place after seeds have naturally worked their way into the soil or have been planted by man.

Thick seed coats must be *scarified;* that is, the seed's surface must first be etched to make minute openings through which water can cross the barrier that the coat presents. In nature, soil fungi and bacteria slowly decompose seed coats; or a more rapid method involves moderate grinding by coarse, shifting soil particles during heavy rains.

Scarification of hard seed coats may also take place during passage of seeds through bird crops or the strongly acidic stomachs of large animals. For some seeds, such a mode of scarification is so essential to germination that they are packaged in colorful, nutritious fruits to attract an animal's attention and reward it with food for having swallowed the seeds. A second benefit seeds derive from such an unlikely relationship is that they receive widespread dispersal. A bird can fly a considerable distance in the time it takes for a seed to pass through its intestines and, at journey's end, the seed is deposited with some fertilizer that may prove useful in getting the new plant started.

Gardeners who collect seeds from their plants for use the following year, may have to use artificial scarification methods. The seeds of many lupine species are a case in point. One rather tedious method is to nick each seed with a file or sharp knife or, alternatively, batches of seeds can be soaked for 1–5 minutes in concentrated sulfuric acid. Exercise caution when using acid and be sure to thoroughly rinse the seeds in running water. Another method is to line a small container with coarse sandpaper, rough side facing inward, put on the lid and vigorously shake the seeds until the coats are well scratched.

Food-Storage Structures and the Embryo

Bean seeds have thin coats, easily peeled off after being soaked for a couple of hours. The bulk of these seeds is occupied by two, kidney-shaped, food-storage structures called *cotyledons,* or seed leaves. (The Greek word *kotyledon* means "cup-like hollow" or concave, as some cotyledons are.) Only when these are carefully pried apart do we find the reason for the seed's being—an *embryo,* a miniature plant waiting for the moment of its birth. The bean embryo shows all of the characteristics of a complete plant, albeit reduced in size: A root, or *radicle* (Latin: *radicula,* small root) as it is called at this stage; a short stem; and a pair of pale leaves that bear evidence of veins within them. The fine details of a plant embryo, like those of an unborn child, are wonderful to behold.

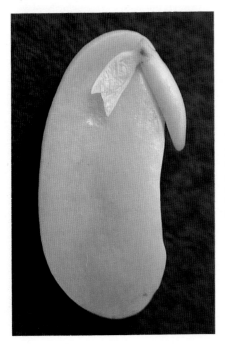

With the seed coat and one cotyledon removed from a bean seed, the embryo is seen pressed against the other cotyledon. Complete with a pair of tiny leaves, a short stem and root, the miniature plant awaits the moment of germination.

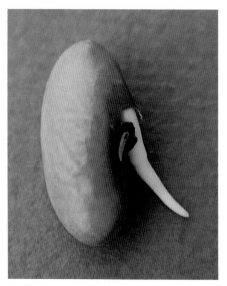

Germination begins with the embryo's rapidly growing root bursting through the seed coat.

While the seedling's stem pushes its way toward the light, the root system branches in several directions from the plant's base. At the soil surface, the hooked stem pulls the cotyledons and, between them, the stem tip out of the soil.

When the cotyledons are free of the soil, the bean's stem straightens and the embryo leaves, present in the seed, begin to expand and turn green.

Close inspection of the bean seedling shows the stem's growing tip between the first pair of leaves. The cotyledons have begun to shrivel as their stored foods are used by the developing plant.

During germination, the embryo is the seed part that grows into a *seedling*, or young plant. At root and stem tips, apical meristems quickly respond to the germination stimulus and launch themselves into the primary growth process described earlier. The embryonic leaves don't expand until they are carried out of the soil by the lengthening stem.

Cotyledons are attached to and, correctly, are a part of the embryo; but their role is entirely different. Rather then growing, they progressively shrink as stored foods are transferred to the seedling. In bean plants, this can be watched since expansion of the lower part of the stem, called the *hypocotyl* (*hypo-*, below + cotyledon) elevates the cotyledons a short distance above the soil surface where, within a few days, they shrivel and eventually drop from the plant; the cotyledons' food supplies have been spent. When plants elevate their cotyledons in this manner, it is called an *epigeous* mode of germination (*epi-*, above; *geo-*, earth). For many seeds, including the pea, the cotyledons remain buried in the ground during *hypogeous* germination (*hypo-*, below).

The seeds of flowering plants contain either one or two cotyledons. Botanists use this characteristic to subdivide the Angiosperms into two major groups—the *dicots* (*di-*, two + cotyledons) and the *monocots* (*mono-*, one). Monocots are believed to be the more "recent" products of plant evolution, compared with dicots, and include grasses, cereal grains (wheat, oats, barley, rice, rye, etc.), sugar cane, bamboo, palms, lilies, iris and orchids. Dicots, the larger group, encompass everything from roses and rhododendrons to ash trees and asters. In addition to cotyledon numbers, other features, to be described later, characterize these two groups of Angiosperms.

A corn grain is actually a seed surrounded by a thin fruit wall to which the seed coat is tightly bonded. The seed contains an embryo, one cotyledon (corn is a monocot) and a second food-storage structure called the *endosperm* (*endo-*, within; *sperma*, seed) that also nourishes the seedling during germination. The soft, white pulp in each grain of fresh corn-on-the-cob is endosperm.

The size of the food-storage structures in a seed determine the maximum depth to which it can be planted and successfully germinate. If, for example, a small seed is set too deeply, the seedling will use the reserve foods, on which it depends, before it reaches the soil surface. Seed packages give specific instructions on depth of planting, but a useful rule of thumb is to bury a seed no deeper than its length. Too shallow is generally better than too deep.

Seed size varies with species of plant, a familiar large seed being that of a coconut. It is the part inside the hard, stony "shell" which is part of the fruit wall. The white, edible "meat" is seed material and the

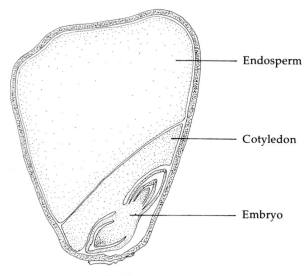

Inside a corn grain.

rich-tasting juice in a fresh coconut is endosperm that has turned to liquid during seed maturation. Gardeners are best acquainted with seeds of less grandiose proportions, including mustard seeds that have been used as figurative symbols of diminished size. Some of the smallest seeds, produced by orchids, are like particles of dust and contain only a rudimentary embryo structure.

Seed Germination

Equipped with everything needed to make a new plant, a seed simply waits until three vital environmental conditions have been met: an ample supply of water, optimum temperature, and situation in a well-aerated soil or other location.

Throughout the months or years of its dormancy, the embryo is held in a state of suspended animation by the dehydrated condition of its cells. Less than 2% of a seed's weight is water, compared with about 95% water in a mature *herbaceous* plant—one that is green and soft textured. It is the low water content that protects a seed against damage at low temperatures. When water freezes it expands and, in cells, ice crystals can tear the protoplasm apart. Yet if a seed becomes completely dry, it loses its viability, its capacity to germinate.

The length of time seeds remain viable varies with species and conditions of storage. Low temperature storage is used in "seed banks" being established throughout the world to protect many plant species against possible extinction due to man's destructive activities. This method of preservation was validated by a report of successful germination of arctic lupine seeds taken from frozen tundra soils and found to be at least 10,000 years old by radioactive-carbon dating methods. Even when stored at room temperature, some seeds remain viable for fairly long periods. Mimosa seeds, discovered in the Natural History Museum in Paris, germinated after 221 years of storage.

A seed, like a dry sponge, can soak up sizeable quantities of water. Water is initially absorbed by a process called *imbibition* in which water molecules fit into spaces between cellulose, proteins and other substances in the dry cell walls and protoplasm. As the cell components absorb more water, they soften and swell, comparable to what happens when dry gelatin granules are soaked in a drop of water; they too imbibe the liquid.

When fully imbibed, most seeds will be about double their original volume. Since seed coats expand to a lesser degree than their contents, the coats split; thus allowing more rapid water uptake by the embryo and cotyledons and also giving them full access to oxygen in the soil atmosphere. Oxygen is needed for the chemistry of what takes place next in the germination process.

Without going into the details of those chemical events, the end result is that large food molecules—starch, protein, fats—in the cotyledons and endosperm, if present, are broken down to smaller, easily transportable units such as sugars and amino acids. Having been sent to the embryo, these substances are used to construct new cells in the seedling's meristems and supply the growth processes with energy. Both plants and animals use exactly the same process (cellular respiration) to extract energy from foods by "burning" them in the presence of oxygen. The gas simply diffuses into plants from their surroundings, including from pores in loose-textured soils. Humans and higher animals· make a more deliberate effort to get oxygen when they breathe.

When sufficient food is available for the embryo, its root pushes into the soil, there to anchor the new plant, take up needed minerals and absorb water by another, more important, method called *osmosis.* The way in which osmosis works will be described in Chapter 8. Soon, it becomes the turn of the embryonic shoot to grow. When it does, the stem tip is curved downward in a hook that pulls the young leaves through the soil, the apical meristem being protected between them.

Throughout its early stages of growth, the seedling is completely dependent upon food supplies from the storage structures of the seed—cotyledons and endosperm. Such a reliance on fully-formed foods from a different source is the hallmark of *heterotrophic nutrition* (*hetero-*, different; *trophic* from the Greek word meaning "nutrition").

Animals and fungi (mushrooms and molds) are heterotrophic organisms. Plants that photosynthesize use *autotrophic nutrition* (*auto-*, self). Only when a seedling's first leaves are lifted into the light does the plant become autotrophic. It is a major switch in behavior from being dependent on foods provided by the seed to becoming an independent organism. Once begun, photosynthesis will produce all the food that the plant will ever again need.

Germination officially ends when the shoot emerges from the soil. Subsequent seedling development includes stem growth, complete expansion of the first leaves—the minute pair, first seen inside the bean seed—and, underground, proliferation of the root system by repeated branching.

Other Germination Requirements

The basic needs of water, ideal temperature and a loose-textured soil to provide oxygen are familiar to anyone who has grown plants from seeds. Perhaps less well known are the special, added requirements of some seeds before they will germinate. Among these are pre-treatments with cold or heat, the need for thorough washing, illumination with red light or, most surprising, being scorched by fire.

After they are shed from their fruits, some seeds will not germinate until they have completed a period of *after-ripening.* This seems to be a time when the embryo matures and all systems in the seed are being readied for the big event. In a batch of seeds from one plant, or groups of plants of the same species, all may not germinate at the same time. As annoying as this may be to the horticulturist, in nature, staggered germination over several months or years is advantageous to a species' survival. If every seed in a population germinated at the same time, the risk of having all seedlings perish in a late frost or. unseasonal drought is increased. While a seed is the stage in a plant's life cycle most resistant to environmental extremes, the seedling is most vulnerable. The system of staggered germination maintains an emergency supply of viable seeds in the soil at all times and is achieved by members of a seed population having different rates of after-ripening or variable rates of scarification due to different seed coat thicknesses.

Another impediment to germination may be the presence of chemicals that inhibit the process. These are generally located in the seed coats and have to be washed away by heavy rains that also wet the soil with enough water to ensure seedling establishment. It is an ingenious way to coordinate germination with periods of adequate rainfall and not have the seeds "fooled" by passing summer showers. Occasionally, germinated seeds may be found inside a grapefruit. But in most fruits, special chemicals or simply a high potassium concentration prevents this from happening. If seeds are to be collected from fleshy fruits, they should be thoroughly washed and dried before planting.

An interesting case of chemical inhibition of germination is one in which a plant may prevent the establishment of others of its own or different species within its growth territory. This effectively eliminates competition for space and resources in short supply. It is called *allelopathy* (literally, "mutual suffering") and is accomplished by saturation of the surrounding soil with chemicals washed out of fallen leaves and twigs from the defensive plant. In a few cases, roots are believed to excrete allelochemicals, which explains the inability of farmers to grow other crops near walnut trees. Allelopathy is a relatively new subject for study but may prove to be important in the development of natural weed killers and in the selection of crops for mixed interplanting.

In order for seedlings to become well established during the most favorable season for growth, seeds of native plants from earth's temperate zones—places where cold winters are normal—must germinate in spring, after the last snows have melted. It would be wasteful for seedlings to start growth late in the year since none could survive winter. To avoid such an outcome, the seeds must be *stratified* before they can germinate; that is, they must be moistened and given an extended period of low temperatures. In nature, this happens in the course of the normal seasonal cycle. Seeds are produced in late summer, are moistened by autumn rains, chilled throughout the winter and are ready to germinate in the mild, sunny days of spring. Seeds possessing this requirement can be artificially stratified by placing them between layers of moist paper, in a refrigerator for a month or two.

When attempting to grow several species of desert wildflowers some years ago, I found that their freshly-collected seeds germinated best when heated in an oven at 120°F (50°C) for one week before planting. Such a harsh treatment is believed to reflect a heat pretreatment requirement for germination in the wild. It is accomplished during the summer months when desert soils and their buried seeds may reach those temperatures. Germination and plant growth are then assured for the winter months, when annual rains are most likely to occur and cooler temperatures prevail.

Seeds of any plant that grows best in bright, direct sunlight are at a disadvantage when they germinate in the shade of other plants. Some sun-loving plants produce seeds that cannot germinate under such unfavorable conditions, responding instead to illumination with only red light. Sunlight is composed of various colors (wavelengths) that we see separated in the bands of a rainbow. When sunlight passes through a leaf, chlorophyll captures the red wavelengths. Below a dense leaf canopy, in a forest of broad-leaved trees in summer, for example, the filtered light is short on red. In a forest of *deciduous* trees (ones that lose their leaves in winter), light sensitive seeds remain dormant until early spring when the leaf canopy has not yet re-grown but temperatures and soil water conditions are favorable for seedling growth. In evergreen tropical rain forests, germination may be delayed for years until the collapse of a large, old tree creates an opening where full sunlight can reach the ground and stimulate the waiting seeds.

Another unusual requirement for germination of some seeds is the need to be scarified by fire. Obviously, such extreme measures apply only to seeds with very thick coats and is most common among species where periodic lightning-caused fires are a part of the balance of nature. In Mediterranean-type climates, including the U.S. southwest, a group of plants is classified under the name "chaparral". These are low-growing shrubs bearing small, leathery leaves, rich in highly flammable resins. Their leaf litter and dry branches make perfect tinder for fast-moving fires, especially on the steep slopes where chaparral normally grows. Seeds from these plants survive the fires with nothing more than a scorching, but sufficient to ready them for water uptake during subsequent rains. The above-ground parts of parent plants are reduced to ash that recycles nutrients back to the soil; nutrients that have been uselessly locked in dead branches for years. Re-growth of chaparral shrubs takes place from underground root crowns. It is more vigorous growth than that which it replaces and more palatable to animals that fled the fires but soon return to start a new life. In the blackened, nutritious soils, seedling growth is rapid and, with the leaf canopy removed, many species of sun-loving plants, especially annuals, occupy formally unfavorable sites, at least temporarily.

Plants favored by man for horticultural and agricultural use share several characteristics; among them are ease and reliability of their seed's germination. Native plant species having unusual germination requirements, such as those described above, are given scant attention and are generally of interest only to professional botanists and the most dedicated garden enthusiasts. But the charm and simple beauty of wildflowers make the extra challenges of growing them in the garden worthwhile; and the inherent capabilities of native plant species to find advantage in even the most adverse conditions, give us optimistic living sermons from which all can benefit. Encouraged by Native Plant Societies in various parts of this and other countries, there is a growing interest in these plants. For those who seek unlimited horizons for inquiry into the wonders of plant life, thousands of wild species, about which little is known, await attention.

Chapter 2

Roots and Shoots.
How Plants Mature

Root Systems

Except for the roots of vegetable crops such as carrots, beets, turnips and radishes, the underground parts of plants receive scant attention from gardeners. It is not so much a matter of roots being "out of sight and out of mind" but, understandably, they do lack the aesthetic appeal of flowers and the attractiveness of leaves. And, perhaps, their pallid, subterranean ways make them slightly incomprehensible. Only when a plant is dug up or a pot-bound plant is transferred to a larger container do most gardeners suddenly become conscious of roots and how much they have grown. Despite our lesser regard for these organs, they are thoroughly worthy of study.

Actually, roots have a certain elegance, largely because of their simplicity. Without sporting eye-catching appendages or putting on a spectacular show, their streamlined structure has been designed by Nature to do three things, and do them well. They anchor the plant in the soil; absorb water and minerals; and store excess food for future needs, underground where animals are least likely to find it.

Roots anchor the plant in one of two ways or, sometimes, by a combination of the two. The first is to occupy a large volume of shallow soil around the plant's base with a *fibrous* (or *diffuse*) *root system,* one consisting of many thin, profusely branched roots. Since these grow relatively close to the soil surface, they effectively control soil erosion; grasses are especially well suited for such a purpose. Fibrous roots capture water as it begins to percolate into the ground and so, must draw their mineral supplies from the surface soil before the nutrients are leached to lower levels.

A *tap root system* sends one or two, rapidly-growing, sparsely-branched roots straight down into the soil to draw from deep water tables and mineral supplies. Tap roots are especially good anchors in shifting soils or windy locations. A few species simultaneously grow both root systems, others adopt one form or the other, depending on soil/water conditions—fibrous roots when the surface soil is moist,

35

A comparison of fibrous and tap root systems.

tap roots when it becomes dry. Because a specific root system is inherited, it is often used as a distinguishing characteristic of a plant family.

Food is stored in both root types, tap roots having the greater storage capacity since they enlarge in diameter as well as length. A carrot is an excellent example of a tap root adapted for food storage and, of course, it is the presence of those foods that makes the root a desirable item in our diets. When the tops of biennial and perennial plants die back in winter or leaves drop and stems become dormant, it is the use of foods stored in their roots that enables such plants to quickly regenerate new foliage the following spring.

The extent of fibrous root growth varies with plant species and availability of soil water, with a tendency to stay near the surface in lightly watered soils. Lawns, for example, should be well soaked at infrequent intervals rather than being given daily sprinkles, this to encourage deep root systems less subject to surface drying and insect attacks on tender root tips. To get an idea of the extent of a well-developed fibrous root system, the roots of an adult rye plant were once counted and measured. It bore approximately 14 million root segments totalling an unbelievable 380 miles (630 kilometers) in length.

Unexpectedly, some large trees have only shallow roots but since they are spreading and matted, form a broad base for support of the trunks. This form of root system is common among trees in tropical rain forests where even the forest giants, as much as 180 ft (60 m) tall, have roots penetrating little more than 3 ft (1 m) into the soil. The advantage gained by such root systems is that they are able to collect nutrients, released from rotting vegetation on the forest floor, before they are washed away by heavy rains in runoff from the shallow soils.

In temperate zones, conifers are generally anchored by deep tap roots that develop large, horizontal branches. Although the roots of most trees grow to moderate lengths, they rarely exceed the height of their uppermost stems. Forest trees, ornamentals, and fruit trees frequently distribute their roots in a wide circle where water-absorbing root tips occupy a "drip zone"—an area beyond the leaf canopy to which rain is channeled from the foliage above. This pattern of root growth should be recalled when irrigating and fertilizing garden trees.

Most horticultural and agronomic plants have relatively shallow roots ranging from 1 to 6 ft (0.3–2 m) in depth. The limited root growth of hybrid roses, for example, simplifies transplanting them either in "root balls" or by "bare root" methods. But among wild plant species, tap roots with lengths of 30–45 ft (10–15 m) are not uncommon. Those of some desert shrubs grow to a vertical depth of over 90 ft (30 m). Cacti, on the other hand, have shallow, spreading fibrous roots to intercept the small amounts of rain penetrating hard, baked desert soil surfaces.

Root Growth

Because the main purpose of roots is to probe the soil for water and minerals at a distance from the center of the plant, primary growth (increase in length) is their most important growth process and apical meristems are their key to success.

Most new cells produced by an apical meristem are laid down behind the growing tip. There they augment the length of the root and when the cells subsequently elongate, the root tip pushes its way through the soil with considerable force. Since a damaged meristem cannot be regenerated, for protection, the meristem also produces cells ahead of itself forming a *root cap*. Root cap cells are readily rubbed off but are quickly replaced from within, much like our skin when it dries and peels from the surface. When root cap cells are ruptured by sharp soil particles, their protoplasm forms a slimy coat lubricating the root tip as it works its way through the soil and around large objects.

Rocks shattered by growing roots, often seen in road cuttings and other excavations, offer impressive testimony to the power of living cells that appear so fragile under a microscope. Their slow, persistent growth was all it took to accomplish such a feat.

Root Hairs and Branches

All of the root's primary growth activities are concentrated in a region about ¼ in. (5 mm) long at its tip. Consequently, water absorption takes place a short way back, in an area where a fuzzy band appears around the root. This band is formed by thousands of projecting *root hairs*. Root hairs are extensions of the outer root cells and increase, several hundred-fold, the organ's absorptive surface area. Their presence increases the rate of water uptake proportionally. Root hairs are easily broken when a root is dug from the soil but can be seen on seedlings (radish is ideal), grown on a moist paper towel in a covered dish for about five days. The width of the root hair zone remains constant. During continued root growth, new hairs form just above the growing tip while old ones, at the top of the group, shrivel and die.

Branching begins in the slightly older root sections, some distance from the tip. Branch roots originate deep inside the parent root (see Chapter 4) and tend to grow at right angles to it, better to explore other regions of the soil around the plant. Each branch is an exact copy of the root that produced it; with an apical meristem, the same methods of growth, a set of root hairs and the capacity to form branches of its own.

Well developed root hairs, as on this radish seedling, absorb most of the water that enters the root. New root hairs are formed toward the growing root tip, while older hairs die back at the top of the root hair zone.

Primary Growth in Stems

A *shoot system* consists of the plant's principal aerial stems, their branches and attached leaves. All have their origins in the stem's apical meristem. A stem's growing tip, its *apical bud,* is much more complex than that of a root, both in structure and activity. It is involved in making the stem grow longer by the normal processes of cell division and elongation; initiates the orderly arrangement of leaves on the stem; and makes provision for the eventual development of branches—all of this within the uppermost 1/10 in. (about 1 mm) of the stem. Details of this remarkable undertaking are readily visualized when an apical bud is viewed through a microscope.

A leaf has been removed to reveal the apical bud of this avocado shoot. Within the pale, tightly wrapped bud, some of the plant's most important growth processes occur.

The apical meristem forms a dome of actively dividing cells. On either side, ear-like lobes represent the first stages of leaf formation which, in this early period of growth, are called *leaf primordia*. These primordia fold over the meristem to protect it against desiccation by sun and wind. At the base of each leaf primordium, a small bulge, an *axillary bud primordium,* is the beginning of a potential branch. Whereas branches form internally in roots, those of stems arise from external buds located in the angle (*axil*) between a leaf and the stem, hence their name. Axillary buds remain dormant until the plant stimulates them to grow. When they do, the primary growth of each branch mimics that of the main stem with an apical bud, leaves, and its own axillary buds for further branching of the shoot system.

When the stem is being formed, it divides into short sections called *nodes* where leaf and axillary bud primordia develop. These alternate with clear stem sections called *internodes* (*inter-*, between). When the stem grows, internodes stretch and spread the leaves (and axillary buds) apart. The result ensures that each leaf is given

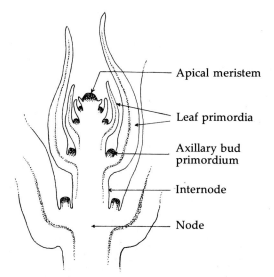

A greatly enlarged view inside an apical bud.

Compare this microscopic view of a coleus stem tip with the accompanying diagram. The apical meristem has established leaf and axillary bud primordia at regular intervals on the sides of the stem.

Arranged in alternating directions at nodes and spaced by the elongating internodes, a coleus' leaves are prepared to intercept as much light as possible.

maximum exposure to sunlight and air circulation and that branches, once formed, are well spaced.

To further reduce crowding of leaves on a stem and their competition for light, the meristem places successive leaf primordia in different directions from those it previously made. Three basic leaf arrangements are shown in the illustration—*alternate, opposite* and *whorled* (arranged in a ring around the stem). Note that, in most plants, pairs of opposite leaves and individual alternate leaves point in changing directions as one traces their arrangement down the stem. To see how effective such an arrangement is in exposing leaves to direct light, look down onto the top of a leafy stem. Despite their numbers, a significant area of each leaf can be seen. To support leaves in such a way that maximum amounts of light can be captured for photosynthesis is the principal function of stems.

Shoot systems are generally capable of producing unlimited numbers of branches. Only a small percentage of their axillary buds grow at any one time; the remainder lie dormant, perhaps for years, to act as points of reserve growth in case apical buds are destroyed by disease, frost, wind or animals. They are also there, ready to grow, after gardeners have satisfied the urge to trim and prune their plants. New growth may appear even from tree stumps, arising from buds long hidden inside the bark.

Some plant groups put all of their growth into one meristem, so to speak; they lack axillary buds. It may seem a chancy strategy, since

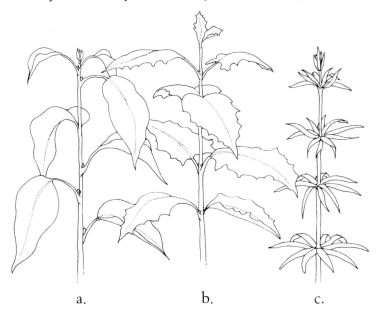

a.   b.   c.

Leaf arrangements: a. alternate, b. opposite, c. whorled.

damage to the single stem tip eliminates all possibility for future growth. Most palms follow this growth pattern, although a few species may sprout new shoots from their bases, but not higher up on established trunks or upright stems.

On an herbaceous stem, an axillary bud (sometimes more than one) is generally visible at each leaf base. To prevent the plant from becoming top heavy, axillary buds normally grow into branches toward the base of the shoot system, thereby maintaining a low center of gravity. But the soft structure of herbaceous stems limits the number of branches, leaves and flowers they can support. Only when Nature introduced wood into the stems of so-called "woody" species was supportive capacity improved.

Axillary buds—
the stem's potential branches.

Woody perennials generally bearing large numbers of branches are classed as either trees or shrubs. What is the difference? By definition, *trees* have one or a small number of main trunks to support their leaf crowns; *shrubs* are smaller plants with many woody stems branching close to the ground. Obviously, such a distinction is not always clear-cut since large, tree-like shrubs or small, shrub-like trees are common. This ambiguity points to a profound quality of the natural order: All too often, human definitions are made to separate black from white whereas Nature tends to work in shades of gray.

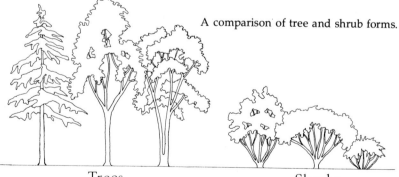

A comparison of tree and shrub forms.

Trees   Shrubs

Development of a Woody Twig

Surrounded as they are by soil, roots do not have the same need to thicken and develop a core of wood. Older parts of roots become woody in most trees, but as a consequence of their age rather than for the supportive role intrinsic to stems. For that reason, root systems contain less wood than the above-ground parts.

In some species, as stems grow longer they simply spread across the soil surface, giving their leaves only minimal support. Such is the case with the runners of some grasses, ivy and other plants we call ground covers. Since the development of wood in these stems is unnecessary, the plant can devote its valuable building materials, food and energy to primary growth processes, the development of leaves and, perhaps, flowers.

Of course, there are advantages in having leaves elevated on upright stems, among which are greater access to direct sunlight and reduced chances of shading. But growth in height demands improved methods for physical support, most often by stem thickening through the introduction of wood during the plant's secondary growth processes. Having built wooden houses, furniture and ships for thousands of years, man has learned that this material is Nature's unsurpassed product for structural and supportive purposes. However, its production by a plant is costly in terms of distribution of the organism's "budget" of available food and energy. What the plant spends on making wood, must be taken from the development of roots, reproductive structures, and leaves where foods are synthesized in the first place. Such allocations of energy budgets between plant parts are of greater interest to scientists than gardeners. But even the most casual consideration gives us another view, from a different perspective, of the enormous complexity of plants that we take so much for granted.

Features of a Woody Twig

At the tips of twigs, where primary growth is taking place, the stem is typically herbaceous. A short way down the stem, the transition into a woody form can be recognized by changes in color, from green to brown, and hardness. Several features of woody twigs are worth noting and are best seen on a "winter twig", one from a deciduous tree following annual leaf loss.

The point on a stem where its color turns to brown is the place where a coat of bark is beginning to form. At first, the bark is smooth and glossy but, with age, it thickens and the surface dries and cracks. The outer bark, called cork, continually flakes or peels off most woody plants but is replaced from within.

Scattered bumps that may look like scale insects on the smooth, young bark of a winter twig are actually breathing pores, called *lenticels,* through which gases, including oxygen, pass to and from the living cells of the inner bark. At nodes on the twig are axillary buds and, below them, *leaf scars,* left as reminders of foliage that fell in autumns past. Inside each leaf scar, small dots can be seen—bundles of food- and water-conducting cells that once exchanged these materials between stem and leaf when the two were united. At the time of leaf separation, the entire leaf scar is covered by a thin layer of cork to seal it against fungal invasion and water loss.

Lenticels—breathing pores in the young bark of a woody stem.

Leaf scars remain on a stem after the leaves have fallen. The circles within each scar are the severed ends of food- and water-conducting bundles. Scattered lenticels dot the stem between leaf scars.

The winter twig's growing tip is encased by overlapping *bud scales* to form a tight *apical* (or *terminal*) *bud.* The apical bud begins to form in autumn, at the time of leaf fall, as part of the tree's preparations for entry into its annual period of dormancy. Being able to escape the ravages of winter, the well-protected bud is ready to respond to the first signs of spring.

Terminal bud scales are pushed aside when the new leaves start their vigorous growth. After the scales drop off, several concentric rings of scars are left, recalling the tree's period of winter dormancy. Because one set of *terminal bud scale scars* is left on the stem each spring, their numbers reveal the twig's age.

Evergreen trees and shrubs do not enter winter dormancy and, therefore, lack the bud scale scars' chronological record. Leaf scars may be found on evergreens' stems but usually toward their base

A dormant bud is encased in glossy bud scales for the winter's duration.

The new growth of spring pushes aside the bud scales.

Rings of terminal bud scale scars indicate where the stem suspended growth for a winter season.

where leaves have dropped off as a result of normal aging processes rather than the coordinated event of an annual leaf fall.

Leaves

None of the wonders of our modern, technological age can match the miraculous awakening of a tree from winter's somber sleep, to put on its spring attire of rich, green, fresh new foliage. Such spectacles of regenerative power are the exclusive domain of Nature's greatest creations.

Leaves are engineering marvels. Their prime purpose is to capture light, the energy source for food manufacture in photosynthesis (see Chapter 8). This is typically fulfilled by the expansion of sheet-like *blades* which must be thin and translucent to allow light to penetrate to their innermost cells. They must also be held in outstretched positions without the assistance of wood in their construction, since wood is opaque and heavy. The blade is frequently attached to the stem by a leaf stalk, or *petiole,* one of the advantages of which is to rotate the leaf blade to track the sun's changing position during the course of the day. A petiole also provides greater flexibility to the leaf in winds and heavy rains and contributes to the spacing of blades for maximum exposure to direct sunlight.

The descriptive term, *petiolate leaf,* meaning one with a petiole, is contrasted with a *sessile leaf* in which the blade is directly attached to the stem (Latin; *petiolus,* stalk; *sessil,* sitting on).

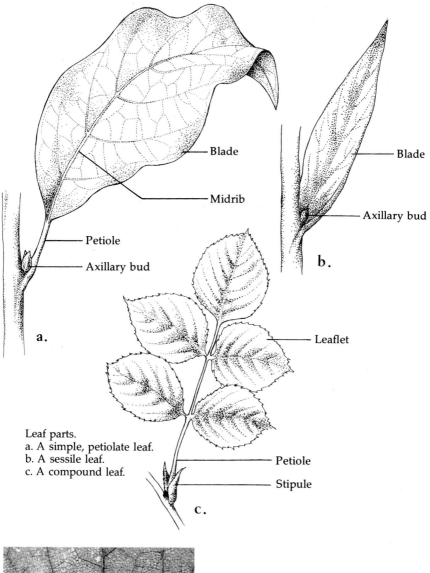

Blade

Midrib

Petiole

Axillary bud

a.

Blade

Axillary bud

b.

Leaflet

Leaf parts.
a. A simple, petiolate leaf.
b. A sessile leaf.
c. A compound leaf.

Petiole

Stipule

c.

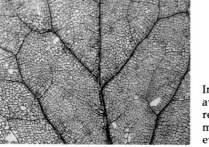

In a "cleared leaf" the soft tissues have rotted away, leaving the intricate network of veins in a reticulate pattern. Branching from the leaf's midrib, veins carry water and food to and from every part of the blade.

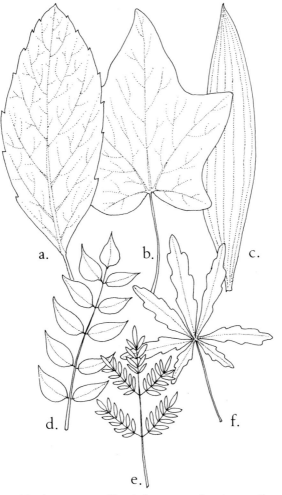

Vein patterns and leaf types. a.–c. Simple leaves: a. pinnate venation, b. palmate venation, c. parallel venation. d.–f. Compound leaves: d. pinnately compound, e. bi-pinnately compound, f. palmately compound.

Leaf blades develop as single units in so-called *simple leaves* or are divided into smaller units, called *leaflets,* in *compound leaves.* Some leaves undergo double or triple compounding in which the leaflets are sub-divided into smaller and smaller segments. The greater the number of divisions, the more "feathery" the appearance of the leaf. A major advantage of compound leaves over simple leaves is that they permit light to pass between the leaflets to lower ranks of leaves. They also tend to be lighter in weight and, therefore, require less support from their stems.

The leaflets of a *pinnately compound leaf* are arranged along a central axis (pinnate means "feather-like"); those of a *palmately compound leaf* arise from one point at the tip of the petiole, like fingers of an out-stretched hand.

Similar descriptions are given to vein patterns within leaf blades: *pinnate venation* and *palmate venation,* in addition to a *parallel* arrangement that is most common in the leaves of monocots (grasses, palms, and iris, for example). Multiple branching of the veins in pinnate and palmate patterns give an overall net-like (*reticulate*) appearance, especially when the softer leaf tissues have rotted away in what is known as a "cleared leaf".

Other features of leaves include overall shape of the blade, shape of the leaf apex and base, and type of leaf margin (lobed, wavy, toothed, and so forth). Samples of leaf forms are shown in the accompanying illustrations.

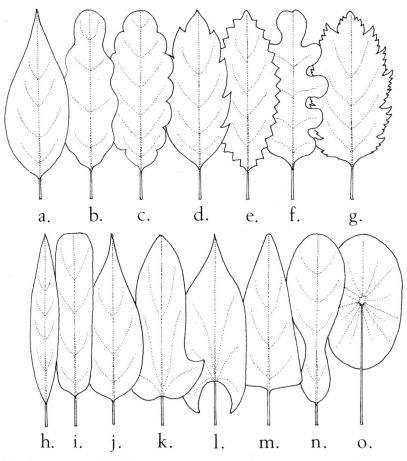

Margin patterns and leaf shapes. Margins: a. entire, b. sinuate, c. crenate, d. serrate, e. dentate, f. lobed, g. double serrate. Shapes: h. linear, i. oblong, j. ovate, k. hastate, l. sagittate, m. deltoid, n. spatulate, o. peltate.

All of these distinctions are used by botanists for taxonomic (classification) purposes, to describe differences between species and groups of related species, called genera. Knowledge of such technicalities is not necessary to appreciate the fact that it is the unlimited diversity of leaf form and color that make most house plants and many garden plants so attractive.

We still do not understand how plants achieve such a miraculous variety of shapes during the course of leaf development—from primordia that all start out looking very much the same to each species' distinctive leaf pattern. A leaf's unique features slowly emerge as blade and petiole expand to their full measure of growth.

Damage to most leaves can never be repaired. If an insect severs a vein, the ends may be sealed over to prevent loss of water, but the hole will not fill with new cells. Grass blades are something of an exception. Everyone who has grown a lawn knows how the grass "keeps growing" after it has been mowed. Few gardeners notice that the pointed tip of each blade, once lost, is not regenerated. Instead, the blades continue growth from near their bases, from what are known as *intercalary meristems*—areas of cell division "inserted" between the blade and the stem. The evolution of intercalary meristems enabled various types of grasses to survive in prairie habitats, in association with herds of grazing animals—deer, antelope, bison and, later, domesticated cows. As long as the animal's teeth simply snip the tops off the grass blades, as a lawn mower does, the leaves continue to grow indefinitely.

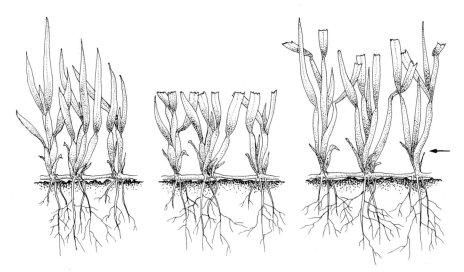

Grass leaves grow from an intercalary meristem, located between the blades and the horizonal stem at the approximate position indicated by the arrow.

The ability to grow is one of the characteristics of living things. Although plants respond best when pampered in our homes and gardens, they grow remarkably well in natural conditions that are rarely ideal. In some habitats, they may be forced to temporarily suspend growth then reanimate themselves as the seasons change. Whatever their secrets for survival, plants have existed for millions of years and, despite man's gloomy predictions for the future of his own race, prospects for the indefinite continuation of a Plant Kingdom look good.

II. ORGANIZATION

Prologue

In the world of government and private business, organization is said to be the key to efficient operation and ultimate success. At a level far more humble yet much older than these institutions, the maxim applies equally well to the maintenance of a well-run garden. In the latter, decisions are made on where to locate plants for optimum growth and how available space can be used to best advantage. Thought must be given to when fertilizers should be applied to best nourish the plants; orders sent in a timely way to procure new seed supplies, roses, bulbs and other temptations from color catalogs. There are pruning schedules and periodic skirmishes with insect pests to be remembered; tools to be returned to places where, later, they will be handy for use. Organization—essential to good gardening but part of the activity's pleasure rather than a rule to be observed.

Plants themselves live highly organized lives. Recall how growth takes place, not in a haphazard way but systematically from meristems. A cell can only divide successfully if it proceeds without variance through the orderly stages of mitosis. And the sequence of events in seed germination is a masterpiece of organization. Throughout every moment of a plant's life, activities of inconceivable complexity take place beneath a deceptive façade of effortlessness and tranquility. Like well-rehearsed professionals, plants give the appearance of ease because their systems of growth, metabolism, adaptation, reproduction have been time-tested and refined over the millions of years of evolution.

When contemplating the intricacies of the human body, its various parts and what each does, it is not possible to separate form from function. For example, a heart is not simply one of many organs but a specialized structure for pumping oxygenated and food-bearing blood to each living cell. Similarly, the parts of a plant must be understood, not simply in terms of their appearance but, more importantly, for what each is intended to do.

In the previous section, the concepts of roots as anchoring and absorbing organs; photosynthesis, the principal function of leaves; and leaves being lifted into the light by supportive stems were developed. Now journey with me inside these familiar parts and, with the aid of a microscope, see how each is precisely constructed from different types of cells. This aspect of botany, one few gardeners have an opportunity to explore, is a part of the scientific specialty of plant *anatomy.*

In the fascinating and beautiful, microscopic world within a plant, organization can be seen at its best. A plant is composed of countless cells having various dimensions, shapes and individual characteristics such as cell wall thickness, being alive or dead at maturity, or having chloroplasts present or absent. These special features determine the specific function of each cell.

Distinctive cell types are not randomly arranged in leaves, roots or stems, but in groups called *tissues.* Collectively, the cells carry out their specialized activity more effectively than do individual cells. For example, transportation of large volumes of water through a plant is carried out by groups of specialized cells, organized into a tissue called *xylem* (from the Greek for "wood" and pronounced "zi-lem"). Water conduction occurs only in an upward direction. Another tissue called *phloem* (Greek; *phloe,* tree bark; pronounced "flo-em") conducts food molecules in opposite directions between leaves and roots. There are tissues devoted to food storage, photosynthesis, support, and protection, in addition to one that simply packs the center of many stems.

Newly divided cells in apical meristems have identical shapes but soon undergo changes resulting in their recognizably different appearances in tissues. How this process of *cell differentiation* takes place is still not understood. Nor is it known how tissues assume the unique patterns characterizing the anatomies of roots, stems and leaves. These are only some of the many unresolved mysteries of developmental processes in plants.

The organizational hierarchy found in a plant actually begins at a much lower level than its component cells—at the invisible level of *atoms.* Of the approximately 100 kinds of atoms or, as more commonly known, *elements,* that compose all matter, only about 20 are employed in plant construction. Elements such as carbon, oxygen, nitrogen and iron fundamentally exist as atoms but are usually combined in various proportions into *molecules.* For example, a water molecule is composed of 2 hydrogen and 1 oxygen atoms—H_2O, as it is written in the chemist's shorthand form. Common table sugar is a crystalline collection of sucrose molecules having 12 carbon, 22 hydrogen and 11 oxygen atoms in each ($C_{12}H_{22}O_{11}$). Proteins, in turn, are even larger molecules made from hundreds of atoms of carbon, hydrogen, oxygen, nitrogen and sulfur. It is important to note that every distinct type of molecule possesses a unique elemental composition and structure.

The next most complex level of organization occurs in living organisms in which countless numbers and types of molecules congregate to form the visible structures of *cells*, including the organelles described in Chapter 1. Cells, in turn, are united into *tissues;* tissues make up the next larger structures of *organs* (roots, stems, leaves, flowers) while they are parts of the whole *organism*, the plant.

Coordination of all living activity is maintained between the component parts at each of these levels, as well as between the different levels. But it is the laws of physics and chemistry governing the interaction between atoms that ultimately control the organism and unify it into a single entity. From air, water and the dust of the earth, atoms unite, albeit temporarily, into living, functioning plants and animals. And when, inevitably, the spark of life is lost, it is back to those primal forms that the elements return.

Chapter 3

Inside Stems

Herbaceous Stems

The soft, flexible tissues of an herbaceous stem are products of primary growth processes and are organized into six clearly defined areas. The outer boundary of the stem is a single layer of cells, called the *epidermis* (Greek: *epi-*, upon; *derma*, skin). The *cuticle* is a layer of waxy *cutin* superimposed on the surface of epidermal cells that reduces evaporative water loss and protects the stem against invasion by molds.

Stems may remain smooth, or *glaucous* (Greek: *glauco*, bluish gray—from the presence of wax), or become hairy, or *pubescent* (Latin for downy), when numerous *epidermal hairs* grow from their surfaces. Epidermal hairs deter attack by small insects, especially when, as in some species, they exude droplets of sticky liquid in which the bugs are snared.

Inside the epidermis, several layers of cells constitute a tissue called the *cortex* (Latin for "shell"). The green color of herbaceous stems results from chloroplasts located in cortex cells. At the stem's center, a large area called *pith* seems to be connected to the cortex due to the similarity of their constituent cells.

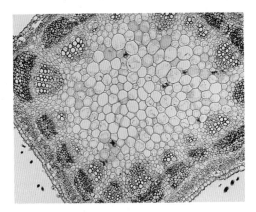

Compare this inside view of a stem with the diagram identifying the various tissues. The most obvious features are the large, central pith and surrounding, red-stained vascular bundles.

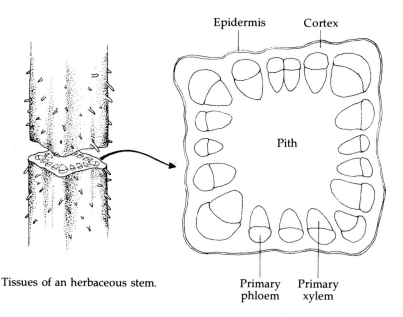

Tissues of an herbaceous stem.

Epidermis   Cortex   Pith   Primary phloem   Primary xylem

The most obvious features of a young stem are bundles of vascular tissues, held in place by the surrounding pith and cortex. The word "vascular" is derived from the Latin *vasculum*, little vessel, and refers to the fact that water, minerals and food molecules are transported through duct-like cells in plants. The inner half of each *vascular bundle* consists of large, water-conducting cells of the xylem tissue; toward the outside are food-conducting phloem cells which are smaller in size. These vital tissues form pipelines of fluid transport, connecting leaves, stems and roots.

The most difficult tissue to identify in a stem is a single row of cells between xylem and phloem in each bundle. This same tissue also separates cortex and pith in the areas between bundles. This is the

A close-up view of two vascular bundles. Phloem tissue in the upper half of each bundle consists of tightly packed food-conducting cells. Rows of large, water-conducting cells constitute the xylem. A row of flat cells between xylem and phloem is the vascular cambium.

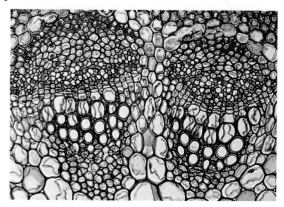

vascular cambium (cambium is Latin for "barrier"), a meristematic tissue whose cells divide laterally (toward the side) and so result in an increase in the stem's diameter during its secondary growth processes. Early in the development of many perennial species, activity of the vascular cambium changes their stems from herbaceous to woody structures. Plant stems that are herbaceous throughout their lives either lack a vascular cambium or have a lateral meristem that remains inactive.

All tissues in an herbaceous stem are established by the apical meristem during primary growth. Thus, they are called *primary tissues.* The food- and water-conducting tissues in the vascular bundles are designated *primary phloem* and *primary xylem* to distinguish them from tissues, formed later by the vascular cambium, called *secondary phloem* and *secondary xylem.*

Primary tissues assume patterns in stems that are almost the same in all species of flowering plants. Keep in mind that the views of tissues shown in the accompanying photographs are only thin slices through their cells. In the three-dimensional structure of a stem, the epidermis, cortex and vascular cambium form concentric cylinders around a central core of pith. Vascular bundles form ribs extending the length of an herbaceous stem and, as such, also function as supportive members; rather like reinforcing steel rods in concrete pillars of modern buildings. But the stem's soft tissues of pith and cortex endow it with great flexibility, allowing it to sway in a breeze without snapping. The arrangement of xylem and phloem in small bundles allows them to branch into leaves and, when axillary buds have grown, into branches of the shoot system.

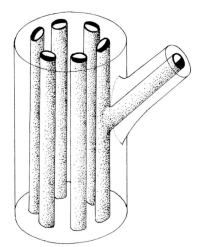

A diagrammatic representation of the vascular bundles in an herbaceous stem; they are held in place by pith and cortex tissues. Both xylem (white) and phloem (black) branch into the attached leaves.

Stem Thickening

For a study in contrasts, compare a tree trunk with an herbaceous stem growing beside it. Although it is easy to forget that the tree, in its infant, seedling stage, was also an herb, its early herbaceous heritage persists at branch tips where new growth is green and soft-textured.

How did the tree trunk undergo changes of such magnitude? The answer lies in the activity of two lateral meristems—vascular cambium, mentioned above, and, after the stem has begun to thicken, a *cork cambium*. The cork cambium is responsible for the formation of *cork*, the outer tissue of the tree's bark.

When cells of vascular cambium divide, they do so in three directions, resulting in different fates for their products. New cells laid down on the inner side of the cambium layer develop thick, lignified

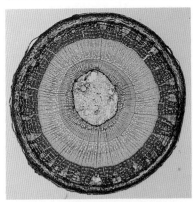

A 2-year old, woody stem. Two concentric rings of secondary xylem (light pink) surround the central pith. Secondary phloem (dark pink) forms a band around the xylem and, on the outside, red-stained cork covers the stem.

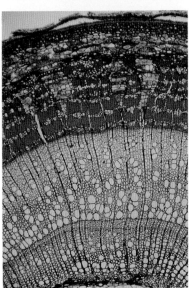

A close-up view of a young, woody stem. A single row of vascular cambium cells forms the dividing line between wood (light area) and bark (dark-stained tissues). The inner bark consists of pink-stained fiber cells alternating with light layers of food-conducting cells in the secondary phloem. Epidermal cells slough off the stem's surface, being replaced by several layers of red-stained cork. The cork cambium is located immediately below the cork in the blue-stained area.

walls and their protoplasm dies. They are thereby destined to become water-conducting cells of the secondary xylem, better known as *wood* cells.

When the vascular cambium divides in an outward direction, secondary phloem is formed. Many secondary phloem cells have thin walls, remain alive and conduct food; others develop thick walls and give physical support to the flimsier food-conducting tissue.

The third direction in which cambium cells divide is sideways— to add more cells to the meristem as it increases its circumference around the growing core of wood.

The vascular cambium forms the dividing line between wood and bark. The inner portion of bark is secondary phloem, the outer part is cork. On a woody stem, cork replaces the epidermis as a protective tissue. Cork is several cell layers thick and is formed by outward divisions of the cork cambium. The latter arises when the epidermis is lost from the original herbaceous stem. When the surface cork of a woody trunk dries, cracks and sloughs off, it is replaced by new tissue from within, thereby maintaining a uniform thickness. The color and pattern of the cork of many woody plants is both an attractive feature for which such plants are grown in landscapes and a distinctive species characteristic.

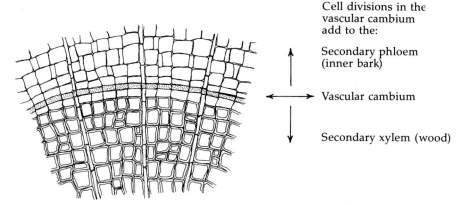

Cell divisions in the vascular cambium add to the:

Secondary phloem (inner bark)

Vascular cambium

Secondary xylem (wood)

Vascular cambium cells (shaded) divide in three directions. The long rows of vascular ray cells also originate in the vascular cambium.

A cross-section through a tree trunk clearly shows the division between wood and the darker bark. The wood (secondary xylem) and the inner bark (secondary phloem) have been laid down by the vascular cambium between the two. Several layers of cork form a rough, protective coat on the tree trunk. Light-colored streaks in the wood are called vascular rays.

Corks used in wine and other bottles are plugs cut from the thick, soft cork of Cork Oak trees, natives of Mediterranean regions. The porous nature of cork tissue plays a significant role in the physiology of a tree for it permits the exchange of gases—oxygen, carbon dioxide, etc.—between the inner, living tissues of the bark and the atmosphere. This same quality makes cork a peerless material for allowing bottled wines to "breathe" during their maturation. In addition, cork cells are naturally water-proofed with a substance called *suberin* (Latin: *suber*, cork) that prevents evaporative losses from the bottles.

Because inward divisions of the vascular cambium are more frequent than divisions in the opposite direction, a tree trunk's wood (secondary xylem) is always thicker than its bark.

Simultaneous with the activities of vascular cambium, cork cambium and, at branch tips, apical meristems are making their contributions to the tree's overall growth. When branches grow, they thicken in the same manner as the main trunk. The base of a branch becomes deeply buried in the trunk as woody tissues grow around it. Knots in lumber are actually slices through these buried branches. Xylem, phloem, the two cambia and cork are all present in, and connected between, a tree trunk and its branches.

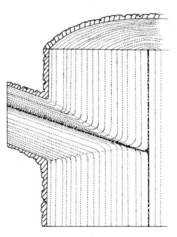

The bark, vascular cambium, and annual rings of wood are connected between a tree trunk and its branches. The base of a branch is deeply buried in the trunk. Knots in lumber are slices through the branch bases.

Branches arise from axillary buds. As they thicken with deposited wood, the bark tissues slowly mature.

Maintenance of this tissue continuity is of critical importance if a gardener is to make a successful graft. Grafting involves the permanent union of a branch (called a *scion;* from an early English word meaning offshoot) taken from one plant, with another plant (a *stock;* old English for stump) that bears roots. In some cases, rose grafting being a typical example, the scion may be no more than an axillary bud attached to a sliver of bark and some wood cells. Alignment of the vascular tissues of stock and scion permits free exchange of nutrients, food and water during the period when the tissues of the two parts fuse together. Grafts can only be made between closely related species of plants; the rejection of incompatible organs is as real in plants as it is in attempted transplants between animals.

Other Features of Wood

The outcome of secondary growth is dramatically revealed in the stump of a felled tree. The obvious line separating bark and wood is the location of the vascular cambium. The thick, woody core is secondary xylem. The most recently formed wood, closest to the cambium, conducts water up the tree trunk and is called *sapwood.* Frequently, sapwood is a lighter color than the inner area, called *heartwood,* the cells of which are plugged with chemical substances and cellular debris. One occasionally sees a large tree still growing after its heartwood has been burned out by a forest fire. This is possible since only sapwood is needed to sustain the tree's life.

Along with materials essential to the plant's well-being, metabolic activities in living cells unavoidably create some waste products. When deposited in the heartwood, such substances discolor the tissue. The rich color of naturally stained heartwood from some tree species makes it particularly desirable for the manufacture of fine furniture. The same waste substances also act as chemical toxins which protect fence posts made of heartwood against fungal attack.

Waste substances are passed into the heartwood from the inner bark by way of narrow rows of cells called *vascular rays,* which are easily recognized in the accompanying photographs. When the vascular cambium adds cells to the secondary vascular tissues, it also adds cells to the progressively lengthening rays. When improperly dried lumber cracks in ray-like patterns in the cross-grain, it does so along lines of structural weakness in wood—the vascular rays.

One of the most interesting features of wood from many species is its pattern of concentric rings, called *annual rings* because each represents one year of growth. Annual ring formation is so regular that the age of a tree may be accurately calculated simply by counting the rings. Because the vascular cambium suspends its activity during

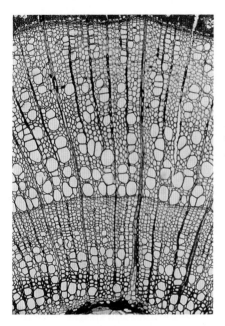

Secondary xylem (wood) is traversed by vascular rays—narrow rows of cells that transport waste materials from living cells in the bark to the inner tissues of the heartwood. Here the wood is divided into two annual rings, in each of which large xylem cells of the springwood merge with smaller cells of summerwood.

fall and winter when the tree becomes dormant, ring growth occurs only during spring and summer. Each annual ring consists of several layers of large xylem cells, called *springwood,* followed by progressively smaller cells—the *summerwood.* To satisfy the demands of rapidly growing, new foliage, larger cells in spring conduct more water which is readily available and abundant in the soil following winter rains and snow melt.

When a log is cut into lumber and other wood products, annual rings, sliced in different directions, form the varied grain patterns that give wood its special aesthetic appeal. Veneers, used in the manufacture of plywood, display the grain's sinuous shapes best of all.

The widths of annual rings vary from year to year, depending on differing annual climatic conditions, especially average annual rainfall. The narrowest rings indicate years of drought, while wide rings were formed in years of abundant moisture. Thus, unsurpassed weather records, dating back thousands of years in Bristlecone Pines, Redwoods and Giant Sequoias are permanently locked into tree trunks. Such records are used by meteorologists to study rainfall cycles. And ring patterns in pieces of wood, matched with those in living trees at archeological sites, offer clues to the mysterious disappearances of early civilizations.

It is nearly inconceivable that a Bristlecone Pine (*Pinus aristata*), still alive, emerged from its seed 1,000 years before Moses led the Children of Israel out of Egypt and that it was "old" when Columbus arrived in the Americas. Barring major climatic changes or natural catastrophe, the future of such living wonders rests entirely in the hands of man. The awesome responsibility of preserving our ancient trees is one which no person can lightly overlook—these majestic, enduring plants demand protection at virtually any cost.

In contrast to the trees of temperate zones, in those of tropical regions the vascular cambium is continuously active since there are no cold winters to bring on dormancy. Their lack of annual rings prevents botanists from learning the exact age of these trees. A variety of studies indicate that many jungle specimens, all Angiosperms, are hundreds of years old but none rival the astonishing longevity of the great Gymnosperms.

Monocot Stems

In Chapter 1, I noted that the flowering plants are divided into two major groups—dicots and monocots—based on the number of cotyledons formed in their seeds. An additional characteristic separating the two groups is the difference in the anatomy of their stems. In a typical monocot, dozens, sometimes hundreds, of vascular bundles are scattered throughout the stem. The bundles are surrounded by uniform masses of thin-walled cells rather than distinct cortex and pith as in the dicot stems, described above. Because vascular and cork cambia are absent, there is no secondary growth leading to the development of wood and bark. Trunk thickening in large monocots, such as palms, simply results from repeated formation of the scattered vascular bundle pattern, formed during a process called *diffuse secondary growth.* Compression and drying of outer cells in a palm trunk create a hard crust that lacks the characteristics of cork in other trees.

The slender trunks of many palm trees reach impressive heights in light of the fact that they are not supported by rigid, wooden cores. This deficiency in forming secondary xylem becomes an asset to native trees in tropical regions where winds of hurricane force shatter trunks and branches made of brittle wood. Pictures of palm trees leaning with the wind attest to their remarkable flexibility.

The stems of other monocots are made even more supple by having hollow centers; bamboo and many grasses are familiar examples. Their stem tissues, including vascular bundles, are supported by large numbers of long, narrow, thick-walled cells called *fibers.* At nodes inside the tube-like stems, the cavity is filled with reinforcing plates to prevent buckling when the stems bend.

A wheat stem may have a ratio of stem length to diameter of 500:1; a giant bamboo 100:1. Compare this with a redwood tree having a trunk height to diameter ratio of only 10:1. The relatively

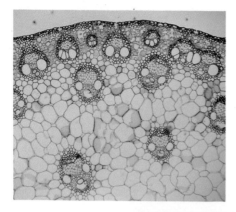

Typical of monocot stems, a corn stem shows the scattered arrangement of its many vascular bundles, embedded in a soft tissue of large, thin-walled cells.

A palm trunk shows the same tissue organization as the corn stem, only much larger. The light dots on the upper surface (lines in side view) are vascular bundles. The absence of hard, rigid wood allows considerable flexibility in heavy winds.

Bamboo stems are masterpieces of structural "engineering". Rings around each stem are nodes where leaves were attached and where internal cross walls support the hollow structure.

light weight, hollow monocot stems can never rival the Gymnosperm in actual height but they effectively elevate their leaves without the expenditure of large amounts of food and energy, required by wood formation.

There are many lessons on superlative organization to he learned from the structure and development of plant stems. The perfectionist's touch, so regularly displayed by Nature, is also seen in the anatomy of roots and leaves, which is the subject of Chapter 4.

Chapter 4

Inside Roots and Leaves

A Root's Primary Tissues

The emergence of distinct tissues, from newly formed cells, occurs a short distance from the root tip. When a thin cross-section is sliced from a young root and examined with a microscope, seven primary tissues may be recognized. On the outside is the *epidermis.* *Root hairs,* discussed in Chapter 2, are extensions of epidermal cells and so extend the surface area of roots, thereby improving the rate of water and nutrient uptake from the soil. In roots, the *cortex* occupies a larger portion of the organ's volume than it does in herbaceous stems.

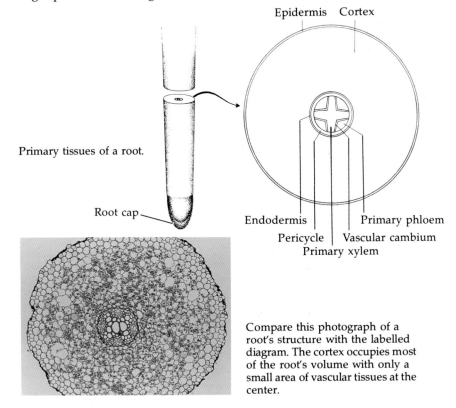

Epidermis   Cortex

Primary tissues of a root.

Root cap

Endodermis   Primary phloem
Pericycle | Vascular cambium
Primary xylem

Compare this photograph of a root's structure with the labelled diagram. The cortex occupies most of the root's volume with only a small area of vascular tissues at the center.

65

Loose packing of cortex cells permit movement of oxygen and water in the intercellular spaces. A plant's underground store of reserve foods is visible in the accompanying photographs in the form of purple-stained starch grains, concentrated in the cortex cells.

Vascular tissues—*primary xylem* and *primary phloem*—occupy the center of a root. Xylem forms an X-shaped configuration; its thick-walled cells are stained dark pink in the photographs. The smaller cells of primary phloem occupy areas between the xylem's "arms". A single row of *vascular cambium* cells, present between the two conducting tissues, is difficult to identify. The vascular tissues are sur-rounded by a row of cells, called the *endodermis.* Its function is to regulate the flow of water across the root and into the xylem.

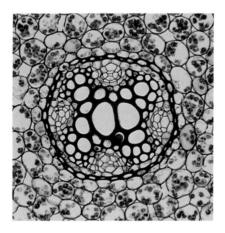

A close-up view of a root's vascular tissues—primary xylem (large, red-stained cells) and 4 groups of primary phloem cells. Vascular cambium is located between the xylem and phloem. The ring of red-stained cells is the endodermis, immediately inside of which is the pericycle—a single row of cells. The large cells of the cortex contain starch grains.

Immediately inside the endodermis is a row of cells, called the *pericycle,* from which branch roots arise. When a branch root develops, it vigorously pushes aside endodermis, cortex and epidermis in its struggle to reach the soil—a rather startling sight, as shown in the photograph. One may wonder why branching begins inside a root and not from buds on the outside, as in stems. The location of the pericycle deep inside a parent root undoubtedly offers protection to this important site of branch formation, better than if it

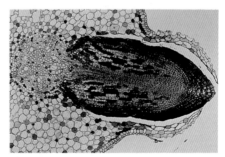

A branch root, arising from the pericycle, pushes aside endodermis, cortex and epidermis cells as it reaches for the soil.

were on or near the root surface. Furthermore, extension of the root's vascular system into a branch root is facilitated by the close proximity of the pericycle and the root's vascular tissues. Branching normally occurs a short distance behind the root tip. When damage occurs to the apical meristem, the pericycle is quickly stimulated to replace the apex with several branch roots. Thus, if a gardener breaks a few root tips while transplanting, the plant soon recovers—but with more than the original number of roots.

At soil level, where root and stem systems unite, transitions between the primary tissue patterns of the two organs must be provided for. The root's central core of vascular tissues divides into the peripheral bundles seen in an herbaceous stem. Epidermis and cortex are continuous between root and stem but, above ground, pith is inserted between the bundles. Because pericycle and endodermis are needed only in roots, they disappear in the region where root turns into stem.

Secondary Growth in Roots

When a tree topples and its larger roots are pulled out of the soil, we can then see how much secondary growth has occurred below ground.

Roots of older perennials thicken in much the same way as trunks and their branches. Vascular cambium lays down wood (secondary xylem) and inner bark (secondary phloem). Cork cambium produces cork on the outside. But the wood in roots is never as extensive as in above-ground parts. Furthermore, it is of little economic value because roots grow in irregular, contorted shapes.

The presence of cork, with waxy suberin in its cells, greatly reduces the capacity for water uptake from the soil by older parts of a root. Most water enters a root in the young portion close to the growing tip, notably in the root hair zone. How roots draw water from the soil and pass it up to the stem and leaves will be discussed in a later chapter.

Cellular Organization in Leaves

The anatomy of a leaf blade is like a sandwich. On either side are layers of epidermis; in the middle are chloroplast-containing cells where photosynthesis takes place. The middle layer is called *meso-phyll* (Greek *meso-*, middle; *phyll*, leaf) and is divided into two parts. Closely-packed, elongated *palisade cells* are arranged directly below the upper epidermis, ready to catch light when it first enters the leaf.

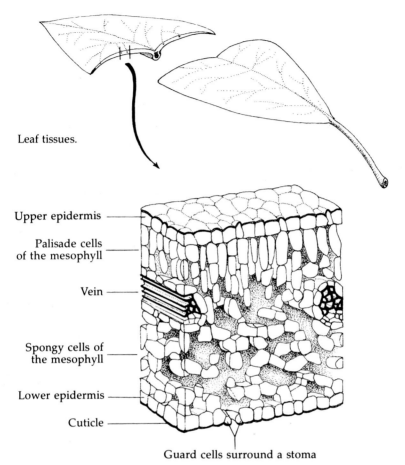

Leaf tissues.

Upper epidermis

Palisade cells
of the mesophyll

Vein

Spongy cells of
the mesophyll

Lower epidermis

Cuticle

Guard cells surround a stoma

The anatomy of a leaf. Tissues may be identified by comparing the photograph with the above diagram.

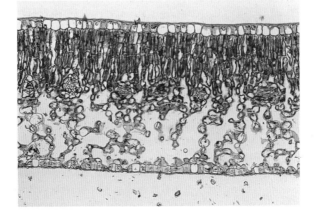

Spongy cells in the mesophyll's lower area are loosely packed to allow gases—carbon dioxide, oxygen and water vapor—to move freely between them. Chloroplasts in the spongy layer capture some of the light that passes through the palisade cells.

Mesophyll cells are, at most, only a few cells removed from veins that weave their way through the leaf blade. Each vein contains a few xylem and phloem cells to supply the mesophyll with water and remove newly-made foods such as sugars.

After spending considerable effort to channel water from the soil to its topmost branches, a plant must restrict water loss from its leaves, or face death from dehydration. The *cuticle* layer on the outer surface of both the upper and lower epidermis provides the barrier that prevents this outcome. Waxy cutin on the leaves of many house plants gives them an attractive, glossy appearance. In contrast, leaves of other species are covered with dense mats of silver-gray, epidermal hairs that help to reduce evaporative water losses from tiny pores in the leaf surface and act as deterrents to insect browsers. They also reflect a portion of the sunlight that strikes a leaf, as does a glossy cuticle—an important safeguard for plants growing in deserts or high on mountains where intense sunlight will damage the structure of chloroplasts.

Most leaves are coated with a waxy cuticle, but not all are as glossy as the *Aucuba* leaf shown here.

Dense mats of epidermal hairs give *Kalanchoe tomentosa's* leaves their gray color and soft texture.

Photosynthesis necessitates that the chloroplasts be supplied with light, water, and carbon dioxide from the atmosphere. The gas enters a leaf through thousands of microscopic pores in its surface—most often, in its lower epidermis. These openings are called *stomata* (Greek: *stoma*, mouth). The location of stomata in the lower epidermis keeps them from becoming plugged with dust that normally collects on upper leaf surfaces. Entry of harmful fungal spores into a leaf is also reduced with stomatal openings placed only in its lower epidermis, for the same reason.

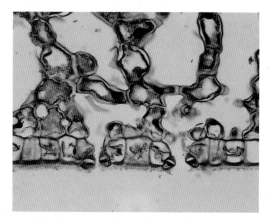

Gas exchange between the leaf's mesophyll tissues and the atmosphere occurs through stomatal pores in the lower epidermis. Two small guard cells border each stoma.

To get some idea of how many of the tiny pores perforate a leaf's surface, consider the following: There are approximately 39,000 stomata per square centimeter of lower epidermis on an apple leaf (1 cm is about ⅜ in.); bean leaves have 25,000/cm^2; orange, 45,000; pumpkin, 27,000. Leaves tending to stand in more of an upright position, such as those of iris, have equal numbers of stomata in both leaf surfaces. Corn (*Zea mays*) has about 6,000/cm^2 in the upper epidermis and 10,000/cm^2 in the lower side. Because the lower side of a water lily leaf is submerged, its stomata are located in the upper epidermis. Water, periodically splashed onto the leaf, washes it free of dust.

Stomatal openings permit gases to enter the leaf. Inadvertently, they are also channels through which water vapor escapes. Periodic stomatal closures are used to regulate such water losses. In most plants, stomata routinely close at night because the absorption of carbon dioxide is unnecessary when photosynthesis is not taking place. Stomata may also close on hot, dry days, in heavy winds or when the soil becomes dry. At such times, photosynthesis may be slowed temporarily but, when water loss exceeds the rate of uptake by roots, it is more important that a plant curb the escape of water than to manufacture foods, of which it usually has a plentiful reserve.

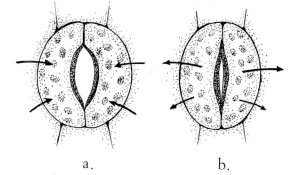

a.   b.

Regulation of stomatal opening. a. A stoma opens when water (arrows) is pumped into the guard cells. The thin, outer wall of each guard cell stretches more than the thick, inner wall. b. When water leaves the guard cells, they relax and the stoma closes.

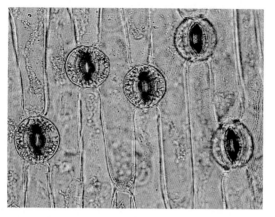

Distinctly different in shape from the leaf's epidermal cells, pairs of guard cells surround each stoma.

Each stoma is bordered by two special cells, called *guard cells,* controlling the size of its opening. The guard cells' inner walls, adjacent to the stoma, are thicker than the outer walls. In a relaxed state, the guard cells lie parallel to each other with no opening between them. But when the leaf pumps water into the guard cells, the thin walls stretch, the shapes of the cells change—curving away from each other—and the stoma opens. Loss of water from the guard cells reverses the process.

The general form and anatomy of leaves are perfectly designed to bring together diverse ingredients for the chemistry of photosynthesis. Most leaf blades are thin enough to permit light to penetrate to the lowest levels of the mesophyll. In thick, fleshy leaves, chloroplasts are located in cells near the surface; the centers of succulent leaves are occupied by large, water-storage cells. Water and dissolved minerals flow through the plant's xylem, connecting roots and stems with leaf petioles, midribs and veins. Carbon dioxide enters a leaf through open stomata, then diffuses into mesophyll cells from the tissue's intercellular spaces where the gas collects. Finally, in the chloroplasts, light and raw materials converge in the process upon which all life depends.

Plant Cell Types

The photographs of root, stem and leaf anatomy in this chapter reveal some obvious differences between cells in the various tissues. Size, shape and cell wall thickness are distinguishing features, made visible by first treating the plant materials with different stains. Cellulose walls are identified with a blue-green dye; lignin appears bright pink.

After careful study of plant tissues, one begins to recognize several basic cell types. Among these are large, thin-walled cells, called *parenchyma,* surrounding primary vascular tissues in roots and stems. The name "parenchyma" comes from the Greek *para,* beside; *enchyma,* an infusion (poured in). Pith and cortex are typical parenchyma tissues. In leaves, palisade and spongy mesophyll tissues are composed of the same kind of cell, their thin walls permitting entry of light to the chloroplasts within. Cells in meristems are a type of parenchyma, but small in size. All newly-formed cells in a plant begin as parenchyma before many differentiate into other cell types, described below. For this reason, parenchyma cells are thought of as the "fundamental" cells of plants.

Investigation of the inherent capacity of parenchyma cells to generate diverse cell types, has led to the concept of *cell totipotency*— such cells are capable of growing into all the tissues and organs constituting complete plants. Under carefully controlled laboratory conditions, in a method called tissue culture, botanists and horticulturists have successfully grown whole plants from small pieces of pith tissue and even from single parenchyma cells. Since only mitotic divisions are involved in the growth of tissue cultures, the resultant offspring, called *clones,* have the same genetic composition as the parent plant. Compare this method of achieving genetic uniformity with conventional methods of cross-breeding that produce *hybrids* having a mixture of genes and, therefore, physical characteristics, inherited from two parents. Cloning techniques possess great potential for the propagation of economically important species, including uniform populations of select, fast-growing, lumber-producing trees. Perhaps it is one of Nature's blessings that the production of human clones appears to be infeasible.

When a plant makes its food-conducting cells, it changes the shape of parenchyma into long, narrow cylinders that, in phloem tissue, are arranged end-to-end in ranks. Their end walls (called *sieve plates*) are pierced with holes, a characteristic giving columns of these cells the name *sieve tubes.* Threads of living cytoplasm pass from cell to cell through the sieve plates. An interesting feature of sieve tubes is that they lack nuclei which are situated in adjoining cells, called *companion cells,* thereby leaving the sieve tubes' cytoplasm free to transport food. This small detail is but one of many illustrations of how sophisticated and specialized plant cells become during the course of their differentiation from parenchyma.

Water-conducting cells are dead at maturity. In flowering plants, these tissues are called *vessels* and have cylindrical shapes and relatively wide diameters. Both the end walls of the cylinders and their protoplasm are absent. Arranged in ranks and columns, vessels form continuous tubes through which water is distributed to all parts of a plant. Their thickened, lignified side walls are perforated with

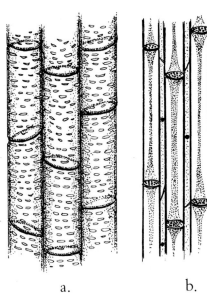

a.   b.

Water- and food-conducting cells. a. Columns of vessel cells in the xylem. Numerous pits perforate the side walls; end walls and proto- plasm are missing. b. Sieve tubes (large cells) and companion cells (small cells) in the phloem. End walls of the sieve tubes are called sieve plates; living cytoplasm con- nects the cells. Companion cells contain the nuclei that control the activities of both cell types.

Five cell types are identifiable in this magnified view of a corn stem's vascular bundle. The 3 large cells are xylem vessels; red stain indicating the presence of lignin in the cell walls. The phloem tissue, with a net-like arrangement of its cells, consists of sieve tubes and smaller companion cells. The vascular bundle is bordered by lignified fiber cells and, sur- rounding the bundle, large parenchyma cells with typically thin, cellulose walls.

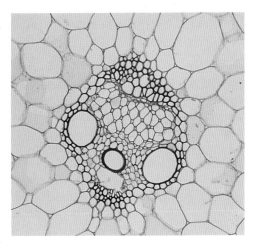

numerous *pits* to permit the escape of water into living cells lining the xylem's water course. In wood, pits allow water to be diverted into adjacent vessels, should blockage occur in any of the cell columns. In Gymnosperms, water-conducting cells called *tracheids* are longer and narrower than vessels but are no less effective in water transport.

Physical support of roots is given by the surrounding soil, whereas shoots must depend on support from their cellular compo- nents. To solve the problem of holding a branch or large leaf in an out- stretched position, yet retain sufficient flexibility to move in a wind, inclusion of *fiber* cells is the ideal solution. Fibers are long and narrow and have thick, lignified walls. Fiber tissues are strong, supple and light weight. Not only are fibers essential to the structural "engineering" of plants, they are of considerable economic importance to mankind. Soft-textured fibers from the stems of flax plants are spun into thread then woven into linen cloth. Coarse fibers from hemp stems and from the leaves of New Zealand flax and agave are used in the manufacture of rope and sackcloth. Fibrous tissues from a wide variety of plant species are also used in the manufacture of baskets, mats, brushes and paper.

Masses of fibers, from which rope is manufactured, project from the frayed end of a New Zealand flax (*Phormium tenax*) leaf. The same leaf, seen under a microscope, reveals ribs of thick-walled fiber cells (stained red) alternating with photosynthetic parenchyma.

Cotton fibers are somewhat different in composition and origin. Their cells walls are composed only of cellulose and each fiber is a long extension of an epidermal cell from the seed coat. Man has used them to make such diverse products as paper, string and clothing. Incidentally, cellulose, extracted from cotton and wood, is used to manufacture some synthetic fibers; most notably, rayon. Other synthetics—nylon, orlon and polyesters—are concoctions made from petroleum products in chemist's laboratories. Since petroleum is derived from the geologic transformation of ancient fossil plants, even synthetic fibers are, indirectly, plant products.

The hardest plant parts are composed of *stone cells* (or *sclereids*). These cells have irregular shapes and extremely thick, lignified walls. They form dense, heavy tissues such as the "pits" of so-called stone fruits—peach, apricot, plum, cherry. Giving more than normal protection to the single seed within, a fruit's stone is the inner part of the fruit wall.

Scientists have reason to take pride in their recent creations, not least of which are extraordinary computers with their tiny, amazingly versatile microchips. But, in the final analysis, all that man has ever invented pales in comparison with the complex organization and functional precision of cells and tissues in even the most lowly, garden weed.

III. ADAPTATION

Prologue

Earth is a restless, ever-changing planet. Its molten core seeps to the surface through giant fissures, spreads across ocean floors and guides entire continents, or parts of them, into new relationships with one another. As if in compensation, sections of earth's mantle sink into the depths and melt under titanic geologic forces. Mountain ranges—Himalayas and Alps—are thrust ever higher by collisions between enormous crustal plates. Volcanoes erupt and cover the land with ash and lava. Ocean boundaries change when land masses rise and subside; lakes and rivers disappear then return when ice sheets periodically creep over the earth and melt in retreat.

Miraculously, through all such upheavals, plants and animals manage to survive, to reproduce and, thereby, to perpetuate genetic lines established when life first began.

Direct descendants of earth's earliest creatures exist today in the form of blue-green algae and bacteria—organisms that have remained virtually unchanged for 3 billion years. Other algae (including seaweeds) and fungi are descended from this ancient ancestral stock, as are mosses and their relatives, the liverworts. Vascular plants, those with vascular tissues, are believed to have evolved from green algae about 400 million years ago; their early species were related to our present-day club mosses and horsetails. Ferns and Gymnosperms came next in the evolutionary succession that culminated in the advent of flowering plants from uncertain origins, approximately 150 million years ago. Over 250,000 species of Angiosperms now populate one or another corner of the earth and at least a few exist in all but its harshest climates.

When, in the course of the never-ending cycle of the seasons, animals find their circumstances increasingly difficult to endure, they are able to move on foot, wing or through water to more congenial locales. Some undertake long migrations in search of better food supplies or desirable places in which to reproduce. Others, to avoid seasonal inclemencies of heat or cold, retreat underground for long or short periods of hibernation. But plants, lacking the ability to move, must survive under the full impact of a changing environment.

A plant, fixed for its lifetime in one location, relies on a panoply of survival strategies, the product of eons of evolution, to meet the contingencies of a fickle and untrustworthy world. Such survival strategies have been acquired through natural selection—the slow, evolutionary process by which only those organisms best able to adapt to environmental change become the progenitors of future generations. Charles Darwin (1809–1882), one of the authors of evolutionary theory, coined the phrase, "survival of the fittest", to describe that outcome.

Natural selection has led to strategies by which plants can survive virtually every conceivable threat to their existence; from cold winters and periods of drought, to attacks by predators and competition with other plants for nutrients and growing space. Many such methods evolved with the development of roots, stems and leaves specially adapted to perform unusual functions. For example, the herbaceous stems of some plants were reduced to compact, frost-resistant, underground bulbs. Cactus stems became the plant's photosynthesizing organs, the leaves having been modified into protective spines. Some roots evolved into climbing structures, while others developed above-ground parts to support heavy tree trunks. The leaves of some species perform the functions of roots by collecting and absorbing water and nutrients. And the most fascinating of all adaptations: Leaves that capture and digest insects as food supplements for species which commonly grow in nutrient-poor soils.

Other survival strategies are found at a molecular level, in the form of natural pesticides—chemical substances synthesized by plants to repel animal browsers and fungal invaders. Some products simply impart strong odors or unpleasant tastes that deter predators, while others kill large or small invasive organisms, so bold as to ingest poisonous plant parts. Such chemicals became more diversified when advanced plant groups, especially flowering plants, evolved. The complex biochemistry of the Angiosperms may well be the basis for their remarkable ability to share diverse habitats with an even greater diversity of animals, including almost a million species of insects.

Although individual plants are unable to migrate, in the course of time, species do move to new locations through the widespread dissemination of spores and seeds. When those reproductive units germinate and plants grow in places far removed from their parents, some of their progeny, in turn, become established even farther from the species' center of origin. A species' distributional range continues to expand until stopped by climatic or other factors antagonistic to growth and reproduction, and to which members of the species are poorly adapted. Formidable barriers such as mountain ranges and oceans, over which transport of most seeds and spores is difficult, may put a temporary halt to dispersion.

Over millions of years, our planet's flora has been in a state of flux. In interglacial periods, when ice sheets released their grip over large portions of the northern hemisphere, plants re-claimed the uncovered land. Pioneer species making the long, slow journey northward, were descendants of ice age survivors in regions closer to the equator, and in whose ranks certain members had a capacity to adapt to climates at higher latitudes.

The inevitable occurrence of genetic variation within a species endows a few individuals with a changed fitness to occupy environments that exclude the bulk of the population. Having thus become geographically isolated from the main body of the species, such splinter groups ultimately give rise to new species, especially when long-standing genetic isolation leads to morphological change. The final stage in the delineation of a new species occurs when its members are no longer able to inter-breed with plants from the ancestral, related, or any other species.

Geographic isolation, favoring speciation, is also believed to take place when cataclysmic geologic events split plant groups into small populations; and when, by chance, a few seeds or spores cross mountains or large bodies of water, borne by migrating animals, high altitude winds or ocean currents. The unique and diverse floras of mid-ocean archipelagos give support to the latter hypothesis and, indeed, it was discovery of the Galapagos Islands' novel flora (as well as its fauna) that gave Charles Darwin a vital clue to the riddle of the origin of species.

Since evolutionary theory was first conceived, scientists have accumulated other evidence to support the idea that, with time, plants and animals undergo significant change. The biological disciplines of genetics (the study of inheritance) and cytology (cell structure), which developed after Darwin's time, have shed light on how genetic variation, leading to geographic isolation and subsequent speciation, takes place within cell nuclei. For example, sexual reproduction produces *hybrids*—offspring in which parental traits are randomly sorted into new combinations. It is not uncommon that, compared with their parents, the progeny's genetic mixture endows them with increased vigor (hybrid vigor) to more successfully cope with adverse environmental conditions.

Mutation is a radical mode of genetic change occurring when the chemical structure of genes is permanently reorganized. The sun's ultraviolet light is believed to be one of several, natural mutagenic agents. Most mutations are thought to be lethal. But those that seem to enhance survival and have been transmitted in the gene pools of plants and animals, have profoundly influenced the course of evolution, especially when, through natural selection, they conferred greater adaptive capacities upon their recipients.

Mutations and the outcome of natural hybridizations are among several random events taking place at a chromosomal level. Their randomness points to the important fact that evolution is not a directed process; it does not work toward pre-determined goals. Yet all that life is and ever has been, is the product of this complex system. Although it is impossible to comprehend evolution in its entirety, the splendid things that Nature has wrought with the process are everywhere to be seen.

Chapter 5

Adaptations for Protection

The Garden Habitat

The gardener's selection of individual plants for indoor or outdoor use is done with a number of interests in mind. Most ornamental varieties are chosen for their aesthetic qualities—the visual appeal of particular flowers or leaves, bark patterns, or overall plant shapes. Plants to be used as ground covers and shade trees are chosen because they are fast-growing species. Fruit-bearing and vegetable crops are selected for their nutrient value, ease of cultivation, and the gardener's tastes. A plant collection may include a few specimens for their novelty—conversation pieces among more ordinary plant neighbors. And, all too often, decisions to purchase certain plants are simply based on cost.

Such reasons for choosing plants are founded on the gardener's personal preferences and economic status. They have no relevance to natural selection, the fundamental principle determining the composition of floras in the wild.

A garden is an artificial habitat in the sense that it is the product of human judgement—but no less enjoyable because of its artificiality. Gardens are among mankind's most praiseworthy accomplishments and, large or small, are unique communities in which introduced plants, having evolutionary origins on different continents, are mingled. Under natural circumstances, such geographically mixed congregations would never exist.

Although much of a garden's design is within the gardener's control, one important factor reveals how Nature is the final arbiter in all plant matters. Namely, the inherent suitability of species for both the geographic locale and the part of the garden where one would like to grow them. For example, attempts to raise tropical species outdoors in Canada would be as futile as planning an alpine garden in Florida. Desert plants in Denmark? Yes, but only in hothouses. Shade-loving forest ferns on a warm, sunny patio? Emphatically no. In other words, the gardener's choices are restricted to species whose native habitats

79

bear some resemblance to the garden or greenhouse environment.

Experienced gardeners instinctively recognize such relationships. Reference books provide information on optimum conditions in which to grow most horticultural species. Some authors conveniently assign commonly grown varieties to "climate zones" into which different parts of the world have been divided. Such zones are determined by: Latitude; altitude above sea level; rainfall patterns; known temperature extremes and frost potential, among other variables.

In mute testimony to the persistence of genetic legacies, each species displays optimum growth only within precisely defined environmental limits, established long ago during ancestral evolution. No matter how many generations removed a plant may be from those ancient ties, it is genetically "programmed" to respond to a specific range of temperature and other climatic and soil conditions. It behooves the gardener to be aware of local conditions not meeting those requirements.

More mysteriously, many plant species also respond to seasonal transitions in day length occurring at latitudes north and south of the equator; for example, some species respond to the change from short days of spring to longer days of mid-summer. Known as *photoperiodism,* this unexpected requirement coordinates plant reproductive cycles with seasons most favorable to growth (see Chapter 7).

The inviolate rule of gardening is that fulfillment of photoperiod and other environmental requisites, having been dictated by Nature, take precedence over man's less consequential concerns for appearance, size or plant cost.

Environmental Modification

Thousands of years ago, in parts of Africa, Asia, the Americas, the Middle East and Mediterranean regions, early peoples established centers of agriculture in which selected native plant species were cultivated for local use. At first, only those crops having food value were propagated but, later, species having ornamental appeal gained man's attention.

As a consequence of mankind's migrations, commerce and, in recent times, rapid means of transportation, an unprecedented diversity of produce and plants is now available for us to enjoy. Supermarkets, nurseries, florist's shops, and gardens are veritable melting pots of plants, formerly from many lands. (See the accompanying tables.)

CENTERS OF ORIGIN OF AGRICULTURE, AND SOME OF THEIR FIRST CROPS

China
(Probably the earliest
center of agriculture)
Apricot (*Prunus*)
Cherry (*Prunus*)
Cucumber (*Cucumis*)
Eggplant (*Solanum*)
Orange (*Citrus*)
Peach (*Prunus*)
Soybean (*Glycine*)
Sugarcane (*Saccharum*)
Tea (*Camelia*)
Walnut (*Juglans*)

India, Malaysia & Indonesia
Banana (*Musa*)
Coconut (*Cocos*)
Ginger (*Zingiber*)
Mango (*Mangifera*)
Mustard (*Brassica*)
Orange (*Citrus*)
Radish (*Raphanus*)
Rice (*Oryza*)
Tangerine (*Citrus*)
Yam (*Dioscorea*)

Middle East & Central Asia
Apple (*Malus*)
Cantaloupe (*Cucumis*)
Carrot (*Daucus*)
Cotton (*Gossypium*)
Fig (*Ficus*)
Garlic (*Allium*)
Grape (*Vitis*)
Leek (*Allium*)
Oat (*Avena*)

Pear (*Pyrus*)
Rye (*Secale*)
Spinach (*Spinacia*)
Turnip (*Brassica*)
Wheat (*Triticum*)

Ethiopia
Bean (*Phaseolus*)
Coffee (*Coffea*)
Okra (*Hibiscus*)
Pea (*Pisum*)

Mediterranean
Asparagus (*Asparagus*)
Celery (*Apium*)
Lettuce (*Lactuca*)
Olive (*Olea*)
Onion (*Allium*)
Parsnip (*Pastinaca*)
Rhubarb (*Rheum*)
Turnip (*Brassica*)

Central & South America
Avocado (*Persea*)
Bell pepper (*Capsicum*)
Cashew (*Anacardium*)
Corn (*Zea*)
Lima bean (*Phaseolus*)
Papaya (*Carica*)
Peanut (*Arachis*)
Pineapple (*Ananas*)
Potato (*Solanum*)
Pumpkin (*Cucurbita*)
Sweet potato (*Ipomoea*)
Tomato (*Lycopersicum*)

PLACES OF ORIGIN OF SOME GARDEN PLANTS
(Genera given when different from common name)

China
Camellia
China aster (*Callistephus*)
Chrysanthemum (*Dendranthema*)
Clematis
Day lily (*Hemerocallis*)

Forsythia
Gardenia
Hollyhock (*Althaea*)
Hydrangea
Peony (*Paeonia*)

Japan
Azalea (*Rhododendron*)
Bleeding heart (*Dicentra*)
Japanese iris (*Iris*)
Wisteria

Australia
Acacia
Bottle brush (*Callistemon*)
Strawflower (*Helichrysum*)

Africa
African violet (*Saintpaulia*)
Bird-of-paradise (*Strelitzia*)
Calla lily (*Zantedeschia*)
Freesia
Gladiolus
Impatiens
Lobelia
Nemesia
Pelargonium
Plumbago (*Ceratostigma*)

Mediterranean
Candytuft (*Iberis*)
Carnation (*Dianthus*)
Grape hyacinth (*Muscari*)
Hyacinth (*Hyacinthus*)
Oleander (*Nerium*)
Snapdragon (*Antirrhinum*)
Sweet alyssum (*Alyssum*)
Sweet pea (*Lathyrus*)

Europe
Bellflower (*Campanula*)
Checkered lily (*Fritillaria*)
Crocus
Forget-me-not (*Myosotis*)
Foxglove (*Digitalis*)

Lily-of-the-valley (*Convallaria*)
Pansy (*Viola*)
Polyanthus (*Primula*)
Primrose (*Primula*)
Rose (*Rosa*)
Scabious (*Scabiosa*)
Snowdrop (*Galanthus*)
Stock (*Matthiola*)
Wallflower (*Cheiranthus*)

North America
Black-eyed Susan (*Rudbeckia*)
California poppy (*Eschscholzia*)
Clarkia
Columbine (*Aguilegia*)
Coreopsis
Lupine (*Lupinus*)
Michaelmas daisy (*Aster*)
Penstemon
Phlox
Sunflower (*Helianthus*)

Mexico
Cosmos
Dahlia
Frangipani (*Plumeria*)
Marigold (*Tagetes*)
Poinsettia (*Euphorbia*)
Zinnia

South America
Fuchsia
Gloxinia (*Sinningia*)
Morning-glory (*Ipomoea*)
Nasturtium (*Tropaeolum*)
Petunia
Portulaca
Salpiglossis
Verbena

An essential part of agricultural and horticultural practice is to modify, in part, the field or garden environment to better suit the inherited requirements of selected species. This we do through irrigation, tilling and fertilizing the soil, pest control, and removing competitive weeds. Some former deserts and other inhospitable places have been turned into productive agricultural regions by such methods. And many rough, unfertile patches of land have been so transformed into gardens.

In greenhouses, limited control of climate is also possible, but productive capacity is restricted by space. Under glass, plants may be protected against frost, snow cover, wind damage, intense sunlight, and the dehydrating effects of low relative humidity. And, as a token of man's genius, optimum growing conditions for almost any species may be artificially created in controlled-environment, plant-growth facilities built for research purposes. But therein lies the extent of man's ability to control the elements. Outdoors, plants are pitted against each and every aspect of the environment, the most destructive forces being those most critically testing the organisms' capacities to endure.

Limiting Factors

The progress of a plant's growth is a summation of its responses to separate, but interacting components of the environment in which it is living. The plant may be favored with adequate water and optimum temperatures but be limited in its ability to photosynthesize by inadequate illumination, perhaps because of shading from taller plants or buildings. Another plant may receive full sunlight, plentiful irrigation and sufficient fertilizer, but still not express its growth potential because prevailing temperatures are too high or too low. Even if climatic and soil conditions are ideal, stunting may occur because pathogenic fungi or predatory insects have invaded the plant. Microorganisms and animals are, indeed, environmental factors to be reckoned with. Other life forms, being ordained components of habitats occupied by plants, exercise both beneficial and harmful effects, as do temperature, rainfall, sunlight, etc.

It becomes obvious that, out of a host of interacting environmental factors, only one need challenge a plant's tolerances in order to limit its growth. The greater the number of unfavorable conditions, acting in concert, the more profound the effect. That is why, in nature, where so many variables are at work, plants rarely reach their full potential. Happily, in a garden, one has the opportunity to improve on a few factors limiting plant development and, consequently, to cultivate larger, healthier specimens than usually occur in the wild.

Plants generally die when too many limiting factors overwhelm their physiological capabilities for survival. Or, on the other hand, they may simply succumb to "old age" processes, principally to a genetically-programmed deterioration of cells and tissues, called senescence (Latin, "to grow old"). Once this process has been initiated, even the best care cannot save a plant. In annual species, senescence takes place within one year of growth; in biennials, in the second year. In perennial species, senescence of a localized nature occurs in older organs before they die and are discarded; it takes many years before the process finally consumes the entire organism.

Protection in Extreme Environments

When a plant becomes dormant, it prepares for the approach of seasons when combined adverse environmental conditions are bound to limit growth, or threaten death. Entry into dormancy entails a reduction of physiological activities, to the minimum needed for survival. At that time, the plant may also discard vulnerable parts, such as leaves prone to damage from frost or the effects of drought. Thus, dormant biennial or perennial temperate-zone species are well-prepared to face winter's low temperatures, strong winds, cloudy days and snow cover. Some desert perennials undergo the same dormancy processes to withstand the long, hot, dry months of summer.

Typically, a dormant plant has well-protected meristems—the sites of renewed growth when environmental conditions improve. Vascular and cork cambia are surrounded by cork tissue, which is not only a superior insulator but, since its cells are impregnated with suberin (page 59), also prevents evaporative water loss. Apical meristems, at stem tips and in axillary buds, are encased in layers of bud scales (page 44)—modified leaves, adapted to withstand prolonged periods of cold or dehydration.

Although annual plants die before the arrival of seasonal temperature extremes or drought, their species survive the worst climatic conditions in the form of dormant seeds, the hardiest structures of higher plants. Among lower plants such as mosses, spores are the units of survival. This type of adaptation is called an avoidance strategy, entailing the passage of a small portion of a plant into the dormant state. The complete organism is genetically programmed to exist only during the most favorable period of the year.

The most taxing problem arising from an avoidance strategy is the accomplishment of both vegetative growth and reproduction within the relatively short life span of an annual plant. This is especially true in deserts where the growing season for annual species is only of 2–4 months duration. If the desert is blessed with abundant year-end rains, followed by mild temperatures and plentiful sunlight, conditions are favorable for the completion of life cycles, from seed germination to seed production. The early arrival of summer's heat brings an end to the entire crop of annual plants. Vegetative growth is, of necessity, minimal. Perhaps four or five leaves, a short stem and a single tap root are all that late-germinating annuals have time to develop—sufficient to provide physical support for a few miniature blossoms and the food necessary for seed development. Despite their diminutive size, these short-lived plants, understandably called "ephemerals", are exquisite gems in the desert flora.

Only occasionally do the several environmental conditions, necessary for the growth of annual species, coincide in the desert; but

Turning the normally parched desert into a springtime riot of color, annual wildflowers crowd around the wizened, perennial shrubs.

Perhaps more beautiful for its small size, a daisy-like *Eriophyllum* spp. is one of hundreds of desert ephemeral species that abundant winter rains call forth from long dormant seeds.

when they do, almost overnight, the normally bleak landscape becomes carpeted with multi-colored flowers in one of Nature's most astonishing displays. In less fortunate circumstances, the seeds merely wait in the soil, year after year, perhaps for decades, for the opportunity to fulfill their destinies.

Environmental extremes of a different and opposite nature dictate small plant size among native species of the arctic tundra and high on mountains in the alpine zone, generally above timberline. Most of these plants are perennials. Their low, compact form provides protection against the crushing weight of snow covers in winter and, after the snows have melted, against the impact of strong winds in their exposed habitats. Another advantage to low growth is that leaves and flowers are positioned close to the ground, in a shallow layer of air warmed when the sun's heat is reflected from the soil. In temperatures a few degrees higher than that of the ambient air, the

At timberline, warmed by the summer sun, alpine lupine flourishes in soils left moist by melting snows.

Although reduced in size, alpine species bear the hallmarks of their better-known relatives at lower elevations. An alpine willow-herb (*Epilobium* spp.) enjoys a brief growing season.

development of low-growing plants is favored, and when pollinating insects fly from flower to flower, they enjoy warm havens and plentiful food supplies.

In their short season of development between dormancies, alpine and arctic tundra plants photosynthesize foods that are stored in roots, ready for use late the following spring when growth is resumed as the last snows melt. Yearly production of flowers and seeds are not as critical events in these perennials as they are in annuals, since reproduction can be attempted in subsequent, favorable years when conditions allow completion of the process.

Many arctic and alpine perennials are evergreen, although minimal photosynthesis occurs during winter. Because their growing season is short, the plants can ill afford the time, and expenditure of food reserves, needed to make a completely new set of leaves each spring.

Leaf cells are prevented from freezing by the presence of high sugar concentrations acting as "antifreeze" in the protoplasm. Dissolved sugars and other cellular substances depress the freezing point of water, as do solutions of specially prepared chemicals sold to protect automobile cooling systems in winter. Plants have been ahead of human invention by several million years.

Perennial species in hot, dry deserts face a different set of problems. Some shed their leaves during periods of drought to reduce the loss of water vapor through open stomata (see Chapter 8). Among evergreen species, the leaves tend to be small, both to expose less heat-absorbing surface to the sun and to reduce stomatal numbers. Other leaf modifications include the presence of extra-thick, water-retaining cuticles and mats of epidermal hairs that slow the evaporation of water and reflect some of the intense light striking the leaf surfaces.

Plants in any habitat constantly struggle to adjust to the changing environment. But none are tested by so many potentially destructive factors, in radical seasonal shifts from one environmental extreme to another, as are the native species of deserts and alpine-tundra regions. Perhaps it is their ability to survive earth's harshest climates that makes these hardy species relatively easy to care for in the garden; provided the basic conditions of native habitats are fulfilled. For warm climate desert perennials such as cacti and succulents, abundant light, infrequent watering and absence of prolonged periods of freezing temperatures are important. Alpine species may only be grown in temperate zones where cold winters, plentiful rainfall, and long summer days are assured. The special beauty of alpine and desert plants may be enjoyed at exhibits sponsored by Alpine Plant Societies and organizations of cactus and succulent plant enthusiasts.

Protection against Animals

It is inevitable that photosynthesizing plants should be the target of destruction by animals since, in the ecosystem, the former are primary food-producers and animals are the principal consumers. Obviously, it is disadvantageous for a plant's well-being, especially its leaves and stems, to be too nutritious and inviting to predators. Some fruits attract animals and provide food as a reward for swallowing and dispersing the seeds (page 26), but that is an exception. Injury from any source reduces plant growth and jeopardizes the chance of reproduction—consequences obviously counter-productive to species survival. Thus, species having effective defenses against predatory animals are especially favored by natural selection and, conversely, the most vulnerable run a greater risk of extinction.

Anyone who has carelessly brushed against a rose bush, cactus, or hawthorn tree, knows how well some plants defend themselves. The protective structures these species bear are classified by botanists into three categories, each an adaptation of a familiar plant part.

Thorns are modified short branches, grown from axillary buds, terminating in sharp, hard points; among other species, Hawthorn (*Crataegus* spp.) and Blackthorn (*Prunus spinosa*) are so equipped.

Spines are modified leaves or parts of leaves, such as projections from the margins of blades. For example, some cactus spines are evolutionary remnants of rigid petioles and midribs, well-sharpened for protective purposes. In the absence of leaf blades, photosynthesis occurs in cactus stems. Some cacti are covered with masses of spines to absorb and reflect excessive sunlight, as do dense mats of epidermal hairs. And spine tips act as places where dew may condense, drip to the soil, and provide moisture for the plant.

The cactus' leaves evolved into protective spines, the stem being the organ of photosynthesis.

A thorn is a modified, short branch. Note that this pyracantha thorn has grown from an axillary bud and, at least temporarily, bears a few leaves.

On the leaves of some species, Holly (*Ilex* spp.) being an example, major veins terminate in *marginal spines* at the blade's edge. Other spines may be adapted from accessory leaf parts called *stipules,* located in pairs at the base of petioles. Since thorns and many spines are modified branches or leaves, they develop at nodes on stems.

Correctly called *prickles,* the rose's protective structures are arranged in irregular patterns within internodes. Prickles are short, woody outgrowths, arising from the epidermal tissue of stems, leaves, and some species of fruit. Many prickles are "recurved"—having tips pointing downward. Such a shape effectively hinders the progress of small animals trying to climb a stem to reach the leaves. And, when present on the long stems of climbing roses and brambles, for example, the recurved prickles become hooked on supports, including other branches of the same plant. This secondary, supportive role played by some prickles also applies to recurved thorns; those on a long, climbing stem of bougainvillaea are shown in the photograph below.

Marginal spines on the holly leaf are extensions of the major veins.

Rose prickles are woody, epidermal outgrowths occurring randomly in internodes.

The recurved thorns of bougainvillaea function both as protective structures and supportive "hooks" for the long branches.

Matted epidermal hairs on *pilose* leaves (bearing long hairs—Latin: *pilos,* hair) and *pubescent* leaves (having short hairs—Latin: *pubesc,* downy) undoubtedly offer protection against many small, herbivorous animals, such as caterpillars, by being difficult to eat. The mouth parts of such animals are adapted to chew on the more substantial tissues of leaves and herbaceous stems.

The special epidermal hairs on stems and leaves of *Urtica,* the stinging nettle, protect in a different manner. When these glandular *stinging hairs* are touched, the tops break off, penetrate skin, and inject a chemical that causes a painful rash. Humans and other animals soon learn to avoid such plants.

Protection by Camouflage

Essential to the perpetuation of Gymnosperm and flowering plant species is the survival and ultimate germination of a part of their annual seed crop. But because most seeds are small, easily digested, and rich in stored food, they are highly desirable items in the diets of many types of animals, from ants to birds, rodents and man. In nature, after seeds are shed from their parent plants, they may lie on the soil surface for long periods before becoming buried. Very small seeds have the advantage of being better able to drop into soil crevices as well as being less visible to the keen eyes of predators. Larger seeds may be protected by thick, hard seed coats or simply rely on their dun color to disguise them against the soil background.

There are many examples of cryptic (hidden from view) coloration among animals; chameleons actually change their skin color to match the surroundings. But few plants are camouflaged to protect them against predators. Several cryptically colored plant species are classed under the popular title, "living stones". These belong to the genera *Lithops, Mesembryanthemum,* and *Conophytum*—natives of the rocky deserts of South Africa. Only the top of the living stone's single

Five "living stones" are camouflaged among pebbles in this small section of a succulent garden. Two of the plants bear dry flower remains.

pair of fleshy leaves projects above the soil; their rounded form and speckled, gray color gives them a convincing, rock-like appearance. Light penetrates to chloroplasts, deep inside, through semi-transparent "windows" at the top of each leaf. Botanists believe that the living stones' camouflage is no mere coincidence, but that their uniquely evolved morphology provides a selective advantage for survival in a habitat where succulent plant tissues are at a premium among thirsty animals.

Protection by Ants

One normally thinks of plants defending themselves against animals, but some fascinating examples have been studied of plants harboring colonies of ants as defensive agents. Frequently, such plants provide both shelter and food for their protectors. The resident ants live inside hollow stems, cup-shaped leaves or large, hollow thorns, as is the case with the Mexican Bullhorn Acacia (*Acacia sphaerocephala*). Some species produce a nutritious liquid, from specialized glands, on which the ants feed. Any unusual disturbance of these plants causes the ferocious residents to swarm out of their nests, ready to attack, giving other insects little chance of encroachment on the ants' territory. Throughout the world, several hundred plant species are known to possess this type of protection.

A mutually-beneficial relationship between two completely different species of organisms is called a *symbiosis* (Greek: *sym*, together; *bios*, life). There are countless plant-animal symbioses, including the important dependency that many flowering plants have upon insects, birds, bats, etc. for the distribution of their pollen (see Chapter 9). But none are as unusual as the association between plants and protective ants.

Wound Healing

Epidermis and cork, the surface tissues of plants, act as barriers between a plant's interior and the external environment. Cutin, produced by and superimposed on epidermal cells (page 54), prevents water loss from leaves and herbaceous stems, and bars entry to fungal spores and mycelium—the thread-like, cellular fungus body. Suberin, a substance in the walls of cork cells (page 59), inhibits water loss from woody stems; whereas *tannin*, another chemical present in cork, acts as a natural fungicide and insecticide.

Injury to either the epidermis or cork results in uncontrollable water loss and the formation of openings through which unwelcome

Wounds left when this tree was shorn of its branches are almost completely covered with corky wound tissue.

organisms find ready access to the plant's interior. Thus, rapid wound-healing in plants is as important in the fight against infection as it is in animals.

An opening made in herbaceous tissues is initially sealed by the exposed cells on the wound surface which collapse and die. Subsequent deposition of waxy substances, similar to cutin and suberin, complete the healing process. On young twigs, a cork layer may also bridge the injured area.

Scars on tree trunks, and branches with well-established secondary growth, are first covered with *callus,* a parenchyma tissue arising from the division of cells near the wound surface. Cork then slowly encroaches from the area around the injury. A few years after branches have been trimmed from a tree or shrub, cork development may have completely obliterated all traces of the work. For healing to be effective, it is important that woody branches be cut as close as possible to the supportive trunks since it is difficult for cork to grow over projecting stumps.

The grafting technique that gardeners employ (page 60) unavoidably creates scars. When a graft "takes hold", proliferation of callus tissue establishes the first connection between the stock and scion. Some of the callus cells then differentiate into vascular and cork cambia, uniting those same tissues in the graft partners. Finally, newly-formed cork and secondary xylem and phloem enclose the graft union.

An injured plant cannot escape being contaminated by the countless fungal spores floating about in the atmosphere and settling on plant surfaces. Since the spores quickly germinate among the newly exposed cells of a wound, isolation of infected areas is necessary to protect healthy tissues. In as much as phloem forms an ideal channel through which fungus mycelium can grow, the vascular tissue provides ready access to every part of the plant by way of its connected, food-laden sieve tubes. Injured phloem, therefore, rapidly responds with the formation of a substance called *callose* and a special protein that plugs sieve plate pores (page 72), to seal broken sieve tubes near the wound. Meanwhile, foods are diverted from the injured tissue to functioning phloem in adjacent areas. Another isolating defense system entails discarding infected leaves, thereby transferring the pathogens to the soil where they assist in the decomposition of the leaf litter. Even before an infected leaf separates from its stem, a tannin-containing cork layer forms across the leaf scar (page 44) to secure it against the spread of microorganisms.

In many species, exudates form effective barriers between injured and healthy tissues. For example, most conifers produce a sticky, aromatic fluid called *resin* that oozes from specialized resin canals when they are broken. Resin is formed in all parts of a conifer tree, is insoluble in water and hardens on exposure to air. Although *gums* are different from resins in their chemical composition and are water-soluble, viscous liquids, they also dry to form hard coats on wounds. Gums are commonplace products of several species of woody flowering plants, including *Acacia,* the source of gum arabic.

Latex is a white or colorless exudate produced by several species of Angiosperms, notably members of the Fig family (Moraceae) and Spurge family (Euphorbiaceae), including Poinsettia. Latex (the name derived from Latin for "fluid") contains, among other components, particles of rubber that effectively seal small scars. Latex, resins and some gums are known to have bactericidal, fungicidal and antiherbivore properties.

These plant products have considerable economic importance. Resins are used in the production of turpentine, rosin, lacquer, varnish, and incense. Gums are used as sizing agents, food thickeners, and stabilizers in emulsions—such as chocolate milk where the gum holds chocolate particles in suspension. Latex is the source of natural rubber, almost all of which is obtained from one tropical tree species, *Hevea brasiliensis* (Euphorbiaceae). Chewing gum is manufactured from another type of latex, called chicle, tapped from the bark of the Chicle tree (*Manilkara zapota*), a native of Central America.

Chemical Protection

Fundamentally, evolution takes place at the gene level. Genetic change leads to modifications in the organism's biochemistry which, in turn, are reflected in observable variations in morphological features and physiological response.

There are two parts to the biochemistry of plants. One involves the life-sustaining chemistry of basic metabolism and includes: photosynthesis; the extraction of energy from foods by cellular respiration (page 20); and the construction of cellulose, starch, fats and proteins. Most important among the proteins are those forming the structure of *enzymes*—catalysts that enable the chemistry of cells to function with great efficiency. Biochemical pathways branching from these essential processes lead to the synthesis of countless *secondary products,* including those that function as chemical defenses.

Basic metabolism, the remnant product of early evolution, is essentially the same in all autotrophic plants. And, as a reminder that all living organisms evolved from the same ancestral first cells, several aspects of basic metabolism are common to both plants and animals. Secondary products, on the other hand, are the biochemical markers distinguishing plant species, families, or higher orders in the taxonomic hierarchy. A few of these chemical substances deserve our attention.

The group of secondary compounds called *tannins,* mentioned above, are a diverse group of molecules characterized by their ability to bind with proteins. In such a role, tannins quickly inactivate enzymes and thereby cause cells to die. Tannins occur in many species of higher plants—especially Angiosperms, Gymnosperms and ferns—and are generally absent from lower forms, such as algae and fungi.

Tannins are found in leaves, unripe fruits, bark, heartwood and roots. In living cells, these substances are safely stored in special structures to prevent them from disrupting normal metabolic activities; when such cells are broken, however, the tannins are released. In dead tissues, such as wood and cork, tannins are present in the cell walls.

During fruit ripening, tannin molecules disintegrate and are replaced by increasing amounts of sugar. The presence of tannins in plant tissues is easily detected by the astringent (dry, puckering) sensation that they cause in one's mouth. A green apple, or strong brew of tea, bear convincing evidence of that fact. The dryness is due to salivary proteins being bound together by the tannins and, thereby, reducing their lubricating action. Interestingly, it is that same astringency that lends appeal to such beverages as tea, wine and cocoa.

The protein-binding, enzyme-inactivating capacity of tannins make them superior deterrents to insects and other herbivores, and substances effectively inhibiting fungal and bacterial growth. For example, heartwood's greater resistance to invasive insects and microorganisms, compared with sapwood, is directly attributable to a 5–10 fold difference in tannin content between the two tissues.

The name "tannin" was originally given to those plant extracts responsible for turning raw animal skins into leather during the tanning process—a procedure that has been known since the dawn of civilization. The word "tan" has its origin in the old English word meaning oak bark. Bark from various species of oak, chestnut, pine, spruce and black locust are still among the important sources of commercially-used tannins. When animal skins are soaked in a concentrated tannin solution, the chemical permeates and attaches itself to the protein fibers and gives the leather the same degree of resistance to microorganisms as that of cork.

One of the most interesting groups of secondary plant products is the *alkaloids*. These nitrogenous substances generally possess alkaline-like qualities (hence, their name) and, when introduced into animals, have wide-ranging physiological effects. The function of alkaloids in plants is not known, although they may protect against predators, in part, because of their bitter taste. But there are some insects that protect themselves against larger predators, such as birds, by making alkaloid-containing plants a regular part of their diet. The chemicals are stored, without harm, in the insect's bodies.

Some authors believe that alkaloids are merely end products of randomly evolving biochemical pathways, for which the producer plants have not yet found use. The distribution of alkaloids in the Plant Kingdom is confined to several species of fungi and club mosses, and a limited number of flowering plant families, including: Amarylidaceae (amarylis), Apocyanaceae (dogbane), Berberidaceae (barberry), Fabaceae (pea), Papaveraceae (poppy), Ranunculaceae (buttercup), and Solanaceae (nightshade).

From man's first use of plants in folk-medicine to the development of our present pharmaceutical industry, alkaloid-containing plant species have played a prominent role in human medicine. Extracts of alkaloids are employed as pain relievers, cardiac and respiratory stimulants, muscle relaxants, blood vessel constrictors, cures for malaria, and pupil dilaters used during eye examinations.

Some alkaloids have mild or strongly addictive side effects: Caffeine in coffee and tea; nicotine from tobacco; cocaine from the leaves of the tropical coca plant; morphine from the opium poppy. Heroin, a synthetic derivative of morphine, is an even more powerful narcotic than the parent chemical. Many of the so-called psychedelic drugs are alkaloids, including mescaline from the peyote cactus and psilocybin from a species of "sacred mushroom". Such plants have been used for centuries in the rituals of several American Indian tribes. The hallucinogen LSD, is a chemically modified form of natural lysergic acid—an alkaloid from ergots, which are fungus infections of grasses.

Social problems arising from drug abuse have, unfortunately, given some plant products a bad reputation. But that should not detract from the fact that mankind's welfare has been vastly improved by the many wonderful chemicals that plants alone are able to produce. One fourth of the prescription drugs sold in the United States in the latter part of the 20th century have had their origins in plants.

Alkaloids are among many secondary plant products, known as *phytotoxins,* that are poisonous to animals. A large number of Angiosperm species contain one or more such substances. Some phytotoxins are distributed throughout entire plants; others are found only in specific organs (see table on page 96). It is not explainable why rhubarb petioles may be safely eaten, when their attached leaf blades contain sufficient oxalic acid to cause muscle and kidney damage, coma, even death. Or why roots and shoots of a tomato plant, but not its fruit and seeds, should contain the violently toxic alkaloid, solanine.

All parts of Poison Hemlock (*Conium maculatum*) are charged with the alkaloid, coniine. The most famous victim of this poison was the Greek philosopher Socrates who, having offended the Athenian government, and according to the custom of the times, was forced to drink a hemlock brew. Ricin, one of Nature's most lethal substances, is present throughout Castorbean (*Ricinus communis*), especially in its attractive-looking seeds. Only 1–3 seeds, if eaten, can be fatal to a child; 2–8 can kill an adult. In the preparation of castor oil, the ricin is removed.

Fortunately, many phytotoxins cause vomiting, a reaction that purges them from the body of an animal before their more sinister work is undertaken. Some plant species possess repellent odors as a warning to animals that poisons are present. And, it has been suggested, the purple-black color of some toxic fruits, such as those of Nightshade (*Solanum* spp.), give a clear "Do not eat" signal to birds and other vertebrates.

Mode of Operation of Chemical Protectants

When eaten, most plant poisons inhibit digestive processes or strike directly at the functions of the heart, liver, kidney, or central nervous system. Less harmful species, such as Poison Ivy (*Toxicodendron radicans*), simply produce substances that, on contact, cause skin irritations.

SOME POISONOUS HOUSE AND GARDEN PLANTS

	Toxic part
Amaryllis (*Hippeastrum puniceum*)	bulbs
Anemone (*Anemone tuberosum*)	entire plant
Apple (*Malus sylvestris*)	seeds, leaves
Apricot (*Prunus armeniacea*)	seeds, leaves
Asparagus (*Asparagus officinalis*)	berries
Azalea (*Rhododendron spp*)	entire plant
Buttercup (*Ranunculus spp*)	entire plant
Caladium (*Caladium bicolor*)	entire plant, esp. leaves, tubers
Croton (*Croton spp*)	seeds
Crown-of-thorns (*Euphorbia milii*)	entire plant
Daffodil (*Narcissus pseudonarcissus*)	bulbs
Datura (*Datura spp*)	entire plant, esp. seeds, leaves
Eggplant (*Solanum melongena*)	leaves, stems
Foxglove (*Digitalis spp*)	entire plant
Gloriosa lily (*Gloriosa spp*)	entire plant, esp. tubers
Holly (*Ilex spp*)	berries
Hyacinth (*Hyacinthus orientalis*)	bulbs
Hydrangea (*Hydrangea spp*)	entire plant
Iris (*Iris spp*)	leaves, rhizomes
Ivy (*Hedera helix*)	berries, leaves
Lantana (*Lantana spp*)	entire plant, esp. berries
Larkspur (*Delphinium spp*)	entire plant
Lily-of-the-valley (*Convallaria majalis*)	entire plant
Lobelia (*Lobelia cardinalis*)	entire plant
Lupine (*Lupinus spp*)	entire plant
Mistletoe (*Phoradendron spp*)	entire plant, esp. berries
Monkshood (*Aconitum ssp*)	entire plant, esp. roots, seeds
Morning-glory (*Ipomoea tricolor*)	seeds
Mountain laurel (*Kalmia latifolia*)	entire plant
Narcissus (*Narcissus spp*)	bulbs
Oleander (*Nerium oleander*)	entire plant, esp. leaves
Peach (*Prunus persica*)	leaves, seeds
Philodendron (*Philodendron spp*)	entire plant
Poinsettia (*Euphorbia pulcherrima*)	leaves, stems, milky sap
Potato (*Solanum tuberosum*)	leaves, stems, green tubers, sprouts
Privet (*Ligustrum japonicum*)	leaves, berries
Rhododendron (*Rhododendron spp*)	entire plant
Rhubarb (*Rheum rhabarbarum*)	leaf blades
Sweet pea (*Lathyrus spp*)	entire plant, esp. seeds
Tobacco (*Nicotiana spp*)	entire plant
Tomato (*Lycopersicon lycopersicum*)	leaves, stems
Virginia creeper (*Parthenocissus quinquefolia*)	berries
Wisteria (*Wisteria spp*)	pods, seeds

(Adapted from: Schmutz, E. M., and L. B. Hamilton. *Plants that Poison.* Northland Press, Flagstaff, Arizona. 1979.)

Secondary plant products having a more subtle mode of operation, are those affecting the reproductive behavior and life-cycles of animal predators. For example, some plant substances resemble animal hormones in their molecular structure. When introduced into an animal's diet, they change reproductive cycles in females or cause growth abnormalities and sterility in males. Other plant products disrupt the larva-to-adult metamorphosis of certain insects and, hence, interrupt the completion of life-cycles.

The antibiotic, penicillin, is synthesized by various species of the fungus *Penicillium,* a common, blue-green-colored mold on rotting fruits and present in blue cheese. Penicillin and other antibiotics, both natural and synthetic in origin, destroy microorganisms in one of several ways: They interfere with cell wall formation, affect the functions of cell membranes, or disrupt the synthesis of proteins and other vital cellular substances. Although none of the biochemical products of higher plants function exactly in the manner of antibiotics, there are several substances known to have specific anti-microbial activity. Some of these, called *phytoalexins* (Greek: *phyton.* plant; *alexi,* to ward off), are synthesized only at the site of pathogen invasion. Others are present, at all times, throughout plants, ready to be mobilized to injured tissues.

Other Methods of Defense

A slightly different use of a phytotoxin is found in the popular house plant, *Dieffenbachia,* the leaves and stems of which possess special cells containing needle-like crystals of calcium oxalate, called *raphides.* If eaten, the raphides penetrate mouth and throat tissues causing painful swelling, difficult breathing, and loss of speech—symptoms that may last for a week or more. Understandably, the species' common name is Dumbcane.

The secondary products of some plant species impart unpleasant, acrid odors; stinging sensations in the eyes; or hot, peppery tastes that deter many animal predators—except man, who is forever searching for new, stimulating food flavors.

Plants may find chemical protection in alternate, less costly ways in terms of energy and nutrient utilization, than through the synthesis of noxious substances. Some species simply accumulate minerals from the soil—copper, lead, cadmium, manganese, selenium, and nitrates—that are toxic to animals. Whereas lignin deposits in cell walls of fibrous and woody tissues, primarily used for support, make plant organs both indigestible and coarse in texture.

The most elementary method for protecting leaves and stems is for the plant to deplete them of all food substances, except those needed for immediate use. Nutritionally-poor shoots are, at least, temporarily protected since animals selectively eat plant species or parts having the highest food value, from among those that are available. But, for the success of this protective strategy, the possession of special, underground structures, where reserve foods may be stored, is a necessity. Such adaptations are discussed in the following chapter.

Chapter 6

Adaptations to Fulfill Basic Needs

Competition Between Plants

In Nature's balance, the blessings the environment bestows on living things outweigh the afflictions capable of destroying them. Thanks to such a slightly favorable balance of good fortune, life is sustained on Planet Earth. Among the benefits plants enjoy are light, water, minerals from the soil, and gases from the atmosphere—precious commodities exploited by plants in both ordinary and extraordinary ways.

In return for what they take from the environment, plants contribute to its improvement. Spreading roots stabilize soils against erosion; vast volumes of liberated water vapor cool and moisten the air; leaf canopies offer shade; trees act as windbreaks; and the ground is enriched with decomposing leaf litter. And to the human mind, which responds not only to physical experience, plants bring pleasure by transforming a bleak world of rock, soil, and urban development into a more pleasant place in which to live.

Paradoxically, an abundance of light, water, and minerals, especially when combined with favorable climate, adds to rather than reduces the problems of plant development. Such conditions encourage an abundance of plants, matched against each other in competition for available resources and growing space.

Most garden landscapes are well-stocked with plants. But, unlike natural habitats, species competition is controlled by the gardener's conscientious pruning, thinning and spacing of specimens. Plants are set in such a way as to avoid casting shadows; or grouped to create needed shade. Plentiful and uniformly distributed supplies of water and fertilizer meet the needs of each plant. Below ground, roots may encroach on neighboring plants' growing space, but can be controlled if, in so doing, they cause noticeable harm.

In the wild, competition between species can be fierce—a fact seemingly contradicted by the apparent tranquility of a cool, shady forest. Especially when limited resources are at stake, plants in densely crowded populations are engaged in a life-and-death struggle to obtain their share.

A direct solution to the problem of sharing limited resources is the use of chemical means to eliminate competitors, of the same or unrelated species. Such is the case with allelopathy (page 33), which is only effective in regions of low rainfall where germination-inhibiting chemicals are able to accumulate in the soil around the defending plant.

In communities of mixed species, some inevitably assume dominant roles because of the relatively large space they occupy, greater use of soil water and mineral supplies, and their interception of most of the direct light. But dominant species pay a price for such favors when they receive the full impact of the environment's destructive forces, especially the damaging effect of wind. Sub-dominant species, adapted to existing with a lesser share of community resources, enjoy the protection given by their more vigorous competitors.

Despite being sub-dominant in a community, some species overcome competitive pressures by use of modified stems, leaves, and roots—organs cleverly adapted to reach bright light, absorb and store water in unusual ways, or exploit unconventional sources of minerals. Many such structures are familiar parts of horticultural species.

Reaching Toward the Sun

Outstretched like arms and fingers, stems spread leafy mantles to intercept light, the vital energy source for photosynthesis. Endowed with the strength of abundant secondary tissues, trees reach great heights where leaves are free of shading. Low-growing shrubs and herbs fare poorly in the presence of tall, overhanging competitors, unless they are *shade-tolerant* species; that is, they have the capacity to photosynthesize in low light intensities. Shade-tolerance is a well-established physiological characteristic of many species; so much so, that their exposure to direct sunlight can be fatal. In such species, bright light destroys the intricate, fine structure of chloroplasts and brings an end to photosynthesis.

The case of most *sun-loving* species is equally calamitous when attempting to grow in heavy shade. The seedlings develop spindly stems, lacking the strength to support themselves and the weight of their leaves. The seeds of many sun-loving species have adapted to avoid germination under a leaf canopy by responding to light quality differences between direct sunlight and that which passes through the shading leaves (see page 33). But such a system restricts sun-loving species to places devoid of larger plants, or periods when sur-rounding deciduous trees have lost their foliage.

In an effort to grow the shortest distance to the brightest avail-able light, the stems of some sun-loving species grow away from shade in a horizontal position; in other species, the stems grow ver-tically, using rigid objects as supports.

Spreading Stems

When stems recline on the soil surface or grow below ground, they have no need to spend energy or nutrients on the construction of metabolically-expensive strengthening tissues. Thus, they are able to direct all of their resources into a burst of rapid, primary growth that carries the leaves into more favorable illumination.

Horizontal stems growing above ground are called *runners,* or *stolons* (Latin for shoot); those growing underground are called *rhizomes* (Greek: *rhizoma,* root). From nodes on stolons and rhizomes, roots and upright shoots develop, the latter from axillary buds. Roots arising from the sides of stems, as these do, are called *adventitious roots* (Latin: *adventicius,* coming from outside). In a broad sense, the word applies to any root emerging in an unusual position, such as those that grow on stem or leaf cuttings.

Stolons generally emerge from near the crown of a plant, bend under their own weight, touch the soil, and develop plantlets at their tips. These, in turn, send out more runners, in a step-wise fashion claiming an ever-widening circle of ground. Thus, a strawberry patch may grow from a single, spreading plant by this natural method of vegetative propagation. It is this same stoloniferous habit that makes some ornamental species well suited to use as ground covers.

The proliferation of underground rhizomes is less apparent, but no less effective, in the occupation of large surface areas of soil. As long as favorably lit places are open for encroachment, the relentless advance of perennial stoloniferous and rhizomatous species may continue indefinitely. A gardener who has cleared a bed of such species knows how persistent they are when, for years after his labors, fragments of broken stems send up sprouts.

Another propagative method, found in blackberry and rasp-berry for example, is the development of suckers. A *sucker* is an upright shoot arising from a horizontal root. Because of its unusual place of origin, a sucker is classified as an *adventitious shoot.* A common problem with grafted hybrid roses is the growth of suckers from adventitious buds located below the graft union. Suckers of this type must be removed since they bear the unwanted characteristics of the root stock, and deprive the scion of nutrients and water.

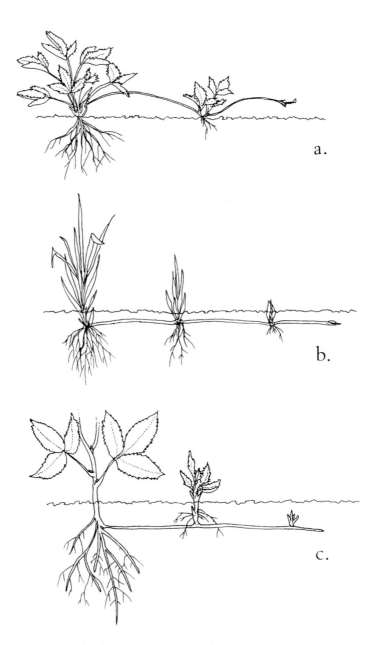

a. A stolon, or runner. b. A rhizome (an underground stem). c. A sucker (an upright shoot, arising from a horizontal root).

Climbing Structures

Most species are unable to elevate leaves high above the ground on thin, herbaceous stems. But not so with climbing vines that make deft use of their specially adapted organs, and the strength of suitable supports, to accomplish such a feat.

The stems of some vines grow in a spiral manner around upright objects such as small trunks of shrubs or saplings, or fence posts and telephone poles. Stems displaying such characteristic growth are called *twiners.* The higher a twining stem grows, the more tightly it hugs its support.

Other species form special grasping organs, called *tendrils,* that are either modified leaf parts or short stems derived from the growth of axillary buds. Tendrils coil around small objects with which they come into contact—the stems of other plants, or garden stakes, fence wires, and string supports. Once anchored, the principal stem grows upward a short distance before sending out more tendrils. *Leaf tendrils* are adapted from leaflets of compound leaves (Sweet Pea, for example), stipules (Green Briar), or petioles (Clematis).

In other species, climbing structures include short branches with adhesive disks at their tips. With such devices, Virginia Creeper (*Parthenocissus* spp.) clings tenaciously to the walls of buildings. Or the climbing stems of ivy, for example, form adventitious roots that penetrate and expand in cracks in tree bark, wooden fences, masonry, etc. With age, some climbing stems become woody and bear heavy leaf loads; but by then, they are so securely anchored it is extremely difficult to separate them from their supports.

Bean plants lift their leaves into the light on coiling twiners.

The terminal leaflets of sweet pea's compound leaves are modified into grasping tendrils.

Virginia Creeper (*Parthenocissus* spp.) clings to a wall with adhesive disks at the tips of specialized branches.

Adventitious, climbing roots growing from the side of an ivy stem are able to penetrate tiny crevices in a supportive surface.

Lianas and Epiphytes

A common sight in tropical rain forests is the long-stemmed, woody, creeping vines, called *lianas* (a French word), festooning tall trees. These species begin life from seeds in the deep shade of the forest floor but rapidly grow to hundreds of feet in length to reach the light. They spread their leaves as the stems work their way through the tree tops, hang suspended from outstretched branches, or become looped to adjacent trees. Liana stems form aerial walkways along which arboreal animals travel great distances; and the vines are an indispensible mode of transport for movie Tarzans.

Other species, occurring in wet, forest habitats from the tropics to temperate zones, are called *epiphytes* (Greek: *epi*, upon; *phytum*, plant)—plants that spend their lives clinging to tree branches. In such elevated locations, their leaves receive optimum illumination. Epiphytes' roots are used more as grasping rather than absorbing organs; few ever reach the soil. Some of the special adaptations epiphytes possess to obtain water and nutrients are discussed below. Many epiphytes grow from windblown seeds or spores, or from seeds deposited on the tree's bark by animals. The most precarious aspect of the epiphytes' existence is that they must share the fate of their supports. Collapse of the host tree generally results in the death of the plant squatters.

Supportive Roots

In the soft, wet soils and crowded conditions of tropical rain forests, some trees have added supports around their bases, thereby gaining a significant competitive edge in the struggle to reach light at ever greater heights. The underground portions of the tree roots are shallow and spreading (page 37), but above ground, huge, wedge-shaped *buttress roots* may extend 10–15 ft (3–5 m) up the sides of the trunk and for an equal or greater distance away from the tree. Buttress roots are common features of several arborescent species of Fig (*Ficus*). Other tropical species, including Screwpine (*Pandanus* spp.), Mangrove (*Rhizophora* spp., *Laguncularia* spp., among others), and Banyan (*Ficus benghalensis*) form *prop roots* (or *stilt roots*) for support. These adventitious roots arise some distance up the trunk, or from branches, from whence they grow downward to the ground. For the same supportive purpose, series of prop roots also grow from nodes around the base of Corn (*Zea mays*) stems when they become top-heavy with their leaf and fruit loads.

Buttress roots support the trunk of a ficus tree.

In the soft, rain-soaked soils of its tropical habitat, Screwpine (*Pandanus* spp.) depends on well-developed prop roots for support.

Special Methods of Water Uptake

Most terrestrial plants absorb water from the soil through their roots. In a few species, water condensed from fog on leaf surfaces is drawn through open stomata. Although little water is absorbed by most leaves, they are instrumental in channelling rain to the roots, either down petioles and stems or directly from the blades to the soil. The leaves of many tropical rain-forest species ensure rapid drying of their surfaces (to discourage growth of fungi, lichens, or mosses), and direct the flow of water, by having smooth blades terminating in pointed *drip tips.* The leaves of the popular house plant, Heart-leaf Philodendron (*Philodendron scandens*), a native of tropical America, illustrate this type of modification.

The bromeliads, an interesting family of plants, mostly from the New World, include many epiphytic species, especially in the genus *Tillandsia.* Living high on the branches of supportive, host trees, their principal source of water is rain collected by the leaves. In some species, cupped *leaf rosettes* form tanks, from which water is absorbed by special cells on the leaf surface. These leafy reservoirs are refuges for many species of small, amphibious animals, as well as breeding pools for mosquito larvae. Animal wastes, combined with rotting vegetable matter and dust collecting in the bromeliad tanks, supply mineral nutrients.

Several other bromeliad species have leaves covered with mats of epidermal hairs that trap and absorb rain water as it streams down the plants. Of these, the best-known species is Spanish Moss (*Tillandsia usneoides*), so-called for its moss-like appearance as it hangs from its host's branches.

Some species of tropical, epiphytic orchids possess *aerial roots* able to collect water vapor from the atmosphere as well as rain. The moisture is absorbed through a soft, white, spongy tissue, called *velamen,* that covers the roots and gives them a silvery appearance.

All but the green tip of an orchid's aerial root is covered with a spongy, white velamen through which water vapor is absorbed from the atmosphere.

Water is stored in a bromeliad's cup-shaped leaf rosette.

Adaptations for Water Storage

To obtain water, perennial species in arid regions either develop a long tap root to reach underground sources or, as is common among many species of cactus, spread horizontal mats of fibrous roots (page 35), just below the soil surface. Although shallow roots become parched and lifeless in the heat of summer, they quickly return to growth and full metabolic activity within hours after rain has soaked the soil. Having taken full advantage of infrequent and unpredictable water supplies, many desert plants survive periods of drought by using water stored in leaves or stems.

The succulent leaves and stems of such genera as *Mesembryan-themum, Sedum, Crassula,* and *Echeveria* contain enlarged water-storage cells capable of supplying the plants' basic needs for many months. Stem succulents such as cacti and cactus-like Euphorbias, sometimes store sufficient moisture to last for years. As much as 95% of the total volume of succulent plants is devoted to water storage. And although most small succulents die by the time half of the stored water is used, some species of cactus have been found to survive a 60–70% moisture loss without damage or significant impairment of physiological functions.

The flat, leaf-like stems of *Opuntia* cactus, called *cladodes,* function both as water-storage organs and light-collecting surfaces. When branching occurs, cladodes are oriented with their flat sides facing in different directions. Thus, in the course of a day, only a portion of the cladodes are subject to the sun's full intensity at one time. In like manner, the ribs and protuberances that are common features of the barrel-shaped and cylindrical stems of other cactus and Euphorbia species, help to shade a part of the plants' surface as the sun changes position. Those same pleats and ribs also enable succulent stems to undergo accordion-like expansion and contraction between times of maximum water storage and periods when reserves are depleted.

The spherical stems of barrel cacti function as organs of water-storage and photosynthesis. With its parallel ribs, only a portion of each stem is exposed to direct light.

Water storage on a grand scale occurs in a few tree species possessing unusually large trunks, specially adapted for that purpose. Baobab (*Adansonia*) trees of the dry savannas of east Africa can store as much as 25,000 gallons (95,000 liters) of water in trunks attaining circumferences of 90 feet (27.5 m). Most plants-people are likely to encounter such species only when visiting botanical gardens.

Underground Food- and Water-storage Organs

Compared with perishable vegetables such as lettuce and celery, onions, potatoes and fresh ginger are ideally suited to long-term storage, with little change in either food value or water content. Classed among several types of modified, underground stems, these organs are adapted to enter dormancy to withstand drought or cold temperatures when, under natural circumstances, they remain in the soil. Fortified with moderate amounts of stored water and abundant reserve food, such stems are ready to grow into full-fledged plants when environmental conditions permit. The high concentrations of food molecules within their cells, acting as "antifreeze" (page 86), prevent frost damage.

An onion is a typical *bulb*. It is a compact shoot, consisting of numerous layers of colorless, fleshy, scale-like leaf bases mounted on a small, disk-shaped stem. The outermost scale leaves are thin and brown, serving to protect the bulb against invasion by soil microorganisms and insects. A central apical bud contains immature foliage leaves that eventually emerge from the bulb. Until the foliage leaves begin photosynthesis, food reserves in the fleshy scale leaves sustain growth.

Axillary buds developing between the bulb's scale leaves, enlarge to become new bulbs. Garlic "cloves" are formed in such a manner. In perennial bulbs (daffodil, for example), flowers develop from axillary buds, leaving the apical bud free to develop foliage leaves in successive years. Whereas in tulip, an annual bulb, the apical bud grows into the flowering stem, thus ending apical growth; so each year's crop of tulip flowers is formed from new bulbs.

Adventitious roots arise from the bulb's flat stem. In some species, when root tips become firmly anchored to the soil, the upper root region contracts by the shortening and thickening of cortex cells. Such *contractile roots* serve to pull the bulb to an appropriate depth in the soil for protection.

Horticulturists propagate bulbous species by digging up the old bulbs and splitting off newly-formed bulbs for individual planting. And the loose scales of lily bulbs may be broken apart and placed under moist, humid conditions where they develop one or more small bulbs at the base of each scale.

An onion bulb consists of layered, fleshy leaf bases attached to a short stem. Numerous adventitious roots arise from the underside of the stem.

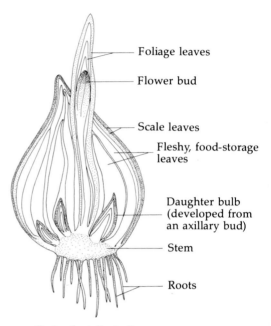

Foliage leaves

Flower bud

Scale leaves

Fleshy, food-storage leaves

Daughter bulb (developed from an axillary bud)

Stem

Roots

Parts of a tulip bulb.

Dry leaf bases

Daughter corm

Cormel

Mother corm (from previous year)

Contractile root

Vegetative reproduction of a gladiolus corm.

Gladiolus corms are short, swollen stems bearing dry leaf bases. Ridges on the corm's surface are nodes.

Although *corms* resemble bulbs in their external appearance, they are dissimilar in structure. Crocus and Gladiolus are cormous species. Corms are short, swollen, underground stems, surrounded by the remains of the previous year's leaf bases; regular roots and contractile roots emerge from the underside of the stem. Parallel lines seen on a defoliated corm are nodes, separated by broader internodes.

Growth of leaves and flowers occurs at the expense of the corm's total food reserves. Following flower formation, one or more new corms develop above the old organ, which eventually withers. In addition, several small corms, called *cormels,* may form around the stem's base; these must grow to their full size before producing flowers.

Another underground, food- and water-storage stem, formed by ginger, bamboo, calla lily, and certain species of iris, for example, is an enlarged, *fleshy rhizome* (page 101). Typical of all stems, rhizomes have internodes, and nodes bearing leaves and axillary buds. From some of the buds, flower stalks develop. Rhizomes are anchored by adventitious roots arising at nodes; several rhizomatous species produce contractile roots.

The white (Irish) potato is another stem adaptation. Called a *stem tuber,* it is the swollen tip of a rhizome, not of the fleshy type. Like rhizomes, stem tubers display the characteristics of the organ from which they are adapted: Axillary buds (the "eyes") are located at nodes; the areas between the eyes are internodes. When the buds develop into shoots, adventitious roots and rhizomes form below them; the rhizomes ultimately bear more tubers. The food reserves of a tuber are completely exhausted by such growth processes.

The fleshy iris rhizome, a modified stem for food and water storage, bears adventitious roots.

A potato tuber is a modified, fleshy stem bearing axillary buds—the "eyes".

Dahlia's root tubers store large amounts of food.

Root tubers (or *tuberous roots*) are formed by such species as Dahlia, Tuberous Begonia, and Sweet Potato. These enlarged organs bear adventitious shoots which, in turn, form adventitious roots that expand to become tubers.

Rhizomes, and stem and root tubers are used extensively by horticulturists as propagative organs, since they can be cut into pieces and planted. However, the gardener must be sure that at least one axillary bud (or adventitious bud in the case of root tubers) is present on each piece. As with all methods of vegetative propagation, the offspring are clones (page 72) of the parent plant and, therefore, possess identical characteristics.

Saprophytes and Parasites

Moist soils are alive with fungi and bacteria. Unlike photosynthesizing plants that produce and store their foods, soil microorganisms absorb food molecules from dead, rotting, organic (plant and animal) matter and use them directly in growth and reproductive processes. Based on their mode of food acquisition, such organisms are classified as *saprophytes* (Greek: *sapros*, rotten; *phyton*, plant). Saprophytes are essential to the decomposition of organic matter and, as a result, contribute to the improvement of soil fertility by releasing mineral nutrients back to the soil, for re-cycling through the roots of higher plants.

Other species of fungi and bacteria play harmful roles when they invade living tissues in search of food. These are the *parasites,* aptly named from the Greek word meaning "to eat at another's table"— which they do in a ruthless manner. Unable to make their own food, they steal from unwilling organisms called "hosts", perhaps facetiously. Many parasites penetrate the host plant's tissues with a special structure called a *haustorium.* In fungal parasites, the haustorium is an extension of the mycelium.

The effects of parasitism range from mild disruption of the host's metabolism to its untimely death. Bacterial parasites cause numerous plant diseases, including: crown gall, cucumber wilt, fireblight, canker, and soft rot of fleshy storage organs. Diseases of fungal origin include: powdery and downy mildews, rusts, smuts, peach leaf curl, apple scab, late blight of potatoes, and damping-off of many species' seedlings. Other plant diseases (lettuce mosaic, tomato spotted wilt, maize dwarf mosaic, for example) are caused by viruses which display a type of parasitic behavior when they invade living cells in order to reproduce. Since viruses are non-cellular in structure, they are not considered living organisms. Experienced gardeners are well aware of the problems of plant disease and the stringent measures necessary to control the pathogens—one of the less pleasant gardening activities.

A few species of flowering plants evolved root-like haustoria with which they pursue a parasitic life style. Some, called *hemiparasites* (Greek; *hemi,* half) or water parasites, principally invade their hosts to obtain water and mineral nutrients. Since they possess chlorophyll, they are able to synthesize most of their required food. Mistletoe (*Phoradendron,* spp.) is one such species attacking many broad-leaved trees and some conifers. Its sticky berries, dispersed by birds, attach to and germinate on the host's bark. Control of mistletoe's growth is difficult since its haustoria penetrate deeply into the host's wood. Witchweed (*Striga* spp.), a hemiparasite attacking the roots of corn, many grasses, and some broad-leaved species, deprives the host of so much water and minerals that its presence is fatal. Death of a host results in death of the parasite; but by then, the interloper has reproduced and its seeds have been spread to other victims.

Angiosperms that are true parasites are completely dependent on their host's food, water and mineral supplies. Among such species is Broomrape (*Orobanche* spp.) that lives attached to the roots of such crop plants as tomato, eggplant, and sunflower. The orange-yellow stems of parasitic Dodder (*Cuscuta,* spp.) grow in tangled webs on the branches of trees, shrubs, and herbaceous hosts such as alfalfa, clover, sugar beet, and some vegetable crops. Removal and destruction of all infected plants is usually the only effective method of eliminating the parasites.

Orange-yellow stems of the parasite, dodder (*Cuscuta* spp.) hang like spider webs on an unwilling host. Seen in cross section under a microscope, the parasite sends haustoria into the vascular tissues of a host's stem (right).

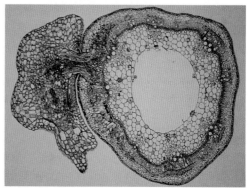

Mycorrhizae and Root Nodules

Mycorrhizae (Greek: *mykes,* fungus; *rhiza,* root) are symbiotic relationships (page 90) between certain beneficial soil fungi and the tender, young roots of many species of higher plants, including: corn, peas, apples, citrus, poplar, oak, rhododendron, birch, pine and other conifers. The fungus penetrates the root tissues, surrounds the roots, and extends into greater volumes of soil than the plant's root hairs (page 38) are able. The nutrients the fungus mycelium taps, especially phosphorus and nitrogen, are used both for its own benefit and that of the host plant. In return, the higher plant supplies the fungus with photosynthesized foods, including sugars.

The beautiful, bright red Snowplant (*Sarcodes sanguinea*), a springtime inhabitant of pine forests in western North America, is an Angiosperm lacking the ability to photosynthesize. It exists through the intercession of a mycorrhizal fungus that transfers food from the roots of nearby trees. The fungus also supplies itself and the other two members of the trio with soil nutrients. Another species having the same, complex symbiotic relationship is the white, waxy-looking Indian Pipe (*Monotropa* spp.).

White mycelium of a mycorrhizal fungus surround the root of a higher plant.

Unable to photosynthesize foods, *Sarcodes sanguinea* (Snowplant) obtains nourishment from the roots of nearby trees by way of a symbiotic, mycorrhizal fungus.

Another symbiosis, mutually beneficial to the two participating organisms, is that of the soil bacterium *Rhizobium* and the young roots of many species of Angiosperms, especially members of the pea family (Fabaceae, formerly Leguminosae).

All living things require a constant supply of nitrogen, especially for the synthesis of cellular proteins. An abundance of the element occurs in the atmosphere in the form of nitrogen gas (N_2), yet few organisms can convert the gas into a form they can use. *Rhizobium* (and several species of blue-green algae) are able to perform such a biochemical trick by a process called *nitrogen-fixation.* The bacteria invade the higher plant's roots, causing them to enlarge in groups of warty *root nodules.* There the microorganisms absorb nitrogen from the soil atmosphere, fix it into valuable ammonium ions (NH_4^+), and pass the product to the root's cells. The host plant's roots supply the bacteria with carbohydrates.

As a consequence of nitrogen-fixation, peas, beans, clover, soybean and alfalfa plants, for example, are especially rich in nitrogenous substances and become a highly desirable mulch when dug into nutrient-poor soils. Clover, soybean and alfalfa are also nutritious species for forage. From an ecological point of view, nitrogen-fixation is a crucial link in the *nitrogen cycle*—a world-wide process in which the nutrient is cycled and recycled between the atmosphere, soil, oceans, and living organisms. As long as the cycle continues, nitrogen will always be available for plants and animals. The same does not hold true for the world's supplies of non-cyclable nutrients, such as phosphorus, potassium, magnesium, and iron. As these elements slowly wash out of the land and into the oceans, they become unrecoverable by natural means and, therefore, unavailable to terrestrial plants.

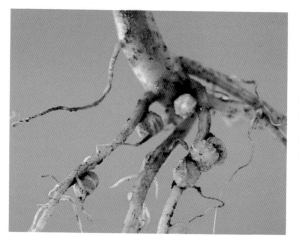

Wart-like nodules on a bean's roots harbor colonies of nitrogen-fixing bacteria in one of Nature's most important symbiotic relationships.

Insectivorous Plants

The most intriguing adaptations in the Plant Kingdom are the specialized leaves of *insectivorous* ("insect eating"), or *carnivorous* ("flesh eating"), species that capture and digest small animals. The evolution of such bizarre behavior, focuses again on the importance of nitrogen to living things, and the devious measures sometimes taken to obtain it. Unless higher plants are favored with a symbiotic partnership with nitrogen-fixing microorganisms, most are unable to exist in habitats with nitrogen-poor soils, including marshes and bogs which are home to insectivorous species.

Like typical Angiosperms, insectivorous plants reproduce by way of flowers, and fruit containing seeds. They support themselves with photosynthesized food materials and are capable of taking nitrogen and other minerals from the soil, when available. But because nitrogen is in short supply in marshes, these remarkable plants rely on their leafy traps. After insects (occasionally, even small birds and amphibians) have been caught, their bodies are digested either by enzymes excreted from glands on the leaf surface, by bacteria present in the leaf traps, or by a combination of the two.

Trapping methods include: *Adhesive traps* in which numerous sticky glands cover the upper leaf surface. Butterwort (*Pinguicula* spp.) and Sundew (*Drosera* spp.) are examples; in the latter, leaf curling, following entrapment, completely engulfs the prey. *Pitfall traps* are common among "pitcher plants" (*Sarracenia* spp. and *Darlingtonia* spp., for example) into whose tubular leaves insects fall and are unable to escape because of slippery surfaces or the presence of sharp, downward-pointed hairs.

Topping the tubular trap (a modified petiole) of a pitcher plant (*Sarracenia purpurea*), the fan-shaped leaf blade is covered with downward-pointing hairs that prevent the escape of captured insects.

Venus' Flytrap's (*Dionaea muscipula*) leaf blades have evolved into ominous traps that snap shut on unwary insect victims.

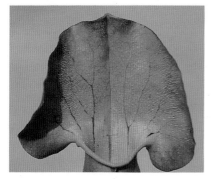

Active traps are best exemplified by the leaves of Venus' Flytrap (*Dionaea muscipula*). The two halves of each blade rapidly close on each other when trigger hairs, projecting from the leaf's inner surface, are touched. The traps may be stimulated to close about 10 times before they will no longer respond. It takes about one day for them to fully re-open. Each trap may catch and digest small insects on three separate occasions. Digestion of a large insect results in the leaf's death.

Earth is blessed with a flora of inconceivable diversity—the outcome of millions of years of natural selection. There is hardly a place to which one or more species is not adapted. And there is hardly an adaptation that does not engender awe. From the simplest forms to the complex Angiosperms and Gymnosperm giants, plants are wondrously attuned to the environment—defending themselves against its onslaughts; taking advantage of good times, while they last. With only roots, stems and leaves to work with, plants modify them, in limitless ways, to survive in environments over which there is no control.

Carnivorous plants, that reverse roles from being consumed to becoming consumers, seem better suited to the realm of science fiction than to life on earth. But what better proof can be offered that, with Nature, innovation has no bounds?

IV. FUNCTIONS

Prologue

What uncanny sense of direction do roots and stems employ when one tunnels through the soil in search of water and nutrients while the other lifts leaves into the light? At predictable times of the year, many species flower in response to specific day and night lengths. Do plants count the passing hours? Why does branching occur when stem tips are removed? Encouraging the gardener to prune plants in order to stimulate more growth seems contrary to all reason. What leads deciduous trees to shed their leaves in fall and awaken from sleep each spring? How do stems bend toward light? And what induces tendrils to coil around vertical supports? There is no end to questions on how plants function, but, unhappily, few positive answers can be offered. Plants are reluctant to share close-guarded secrets.

The study of functions, or *physiology,* deals with the inner workings of organisms. Not content to simply accept observable phenomena, biologists seek underlying reasons for what they see and thereby open new doors of discovery to evermore exciting, invisible worlds.

Plant physiology is built on a solid foundation of information about plant structure, established by such botanical disciplines as morphology, anatomy, and cytology. But when anatomists describe the growth of parenchyma and their differentiation into other types of tissues (page 72), physiologists want to know *how* those changes take place at the molecular level. For centuries, morphologists (and countless gardeners) have observed the sudden appearance of flowers on plants that, for months, or years, have produced only leaves. What hidden biochemical events evoke such an astonishing transformation? With powerful microscopes, cytologists see the miraculous internal organization of cells and their organelles (page 19). For what purpose has Nature designed each intricate part?

The key to many mysteries of how plants function lies in the chemistry of their cells—the very reason why access to the answers is so difficult. Biochemical processes are exceedingly complex, involving thousands of simultaneous chemical reactions in individual cells and tissues. A single sequence of molecular events may hold the clue to how a physiological process takes place. But even with the help of modern, sophisticated instruments, to seek and isolate that sequence is like looking for the proverbial needle in a haystack.

In an age when people have walked on the moon, and lives are saved with transplanted organs, it is humbling to admit that routine functions in "simple" plants still baffle mankind. For example, there is no completely satisfactory explanation for how food molecules move through phloem cells; one of the earliest proposals was made in 1860, yet plant physiologists continue to be puzzled. Or consider cell membranes: They are visible under electron microscopes and, from extensive experimental work, have been observed to be highly selective about which substances they admit to the cytoplasm. But how such selectivity is achieved remains a matter of speculation. Again, it is common knowledge that certain minerals, such as boron and molybdenum, are needed by plants, but it is not clearly understood how they are used. Mankind's perplexity over such matters tells us, perhaps, that plants are not as simple as their appearance suggests.

Much of what we know of plant physiology is based on results from carefully designed experiments. Experimental scientists, whether biologists, chemists or physicists, employ a common systematic approach in their work when they follow the so-called "scientific method".

Research is begun when observations of particular biological or physical phenomena are made, both directly by the investigator and, indirectly, through the accounts of other scientists (and, in some cases, amateurs) in scientific and other publications. Extensive research in the library is an important prerequisite to experimentation. For example, a plant physiologist, interested in the Venus' Flytrap and its rapid leaf closure, may spend months searching the literature to become thoroughly acquainted with previously reported information and opinions on both the specific and related topics; in this case, plant movements in general.

The second stage of the scientific method is the formulation of an hypothesis—a provisional conjecture, based solely on preliminary observations, of how the phenomenon takes place. The hypothesis is then tested by a series of carefully planned experiments. To be of value, such experiments must focus on the specific objective of study by limiting the number of external factors that may influence the outcome; and then must be repeated several times to determine whether comparable results are obtainable. Well-planned experimental design, the accuracy of the techniques employed, and the ability of other scientists to duplicate the work, are crucial to the quality of scientific endeavor.

Experiments may include: Laboratory tests of the plant's responses to various treatments; studies of the organism in its native habitat; microscope examination of cells and tissues; or a combination of these and other methods. The results of each experiment are recorded and, from time to time, evaluated for their contribution to an understanding of the topic under study. From analysis of the accumulated data, other experiments may be undertaken, techniques refined, and different approaches to the problem devised.

Finally, when sufficient, and convincing evidence has been collected for presentation to the scientific community, conclusions are drawn which may, or may not, support the original hypothesis. Regardless of the outcome, the gathered information is of use to other scientists only if it is reported factually and without bias on the part of the investigator. Nowhere, in all human knowledge, must truth be accounted for more rigorously than in the world of science.

Plant physiology is divided into three topical areas: The physiology of growth and development, including chemical and environmental regulation of those processes; systems for the uptake and transport of raw materials from the environment—water, soil nutrients, and gases; the utilization of those materials in photosynthesis, and the channeling of photosynthetic products into cellular metabolism.

The persistent fact that questions about plant functions far outnumber the answers has spurred physiology to become the fastest growing, and most intensely studied subject in botany today. From an overwhelming body of information on the subject, topics of practical interest to the reader have been selected for the following chapters. With an understanding of the basics of plant physiology and a little imagination, investigative gardeners may wish to devise some experiments of their own, as much can be done in the home and garden with improvised methods and equipment. To the adventurous, there are no restraints upon the discovery of how plants function.

Chapter 7

Control of Growth and Development

Growth Responses to Light

At the turn of the 20th century, plant physiologists and biochemists joined forces to search for the molecular controls behind the processes of plant growth and development. For some time, it had been speculated that plants produce special substances, similar to hormones in animals, to regulate such systems. The name "hormone" is derived from a Greek word meaning "to excite". In a broad sense, hormones initiate biochemical activities resulting in observable, physiological responses.

Plant hormones, or *plant growth regulators* as they are sometimes called, proved to be different from animal hormones in chemical structure, mode of synthesis, and function. As part of their endocrine system, higher animals possess glands, organs specialized for the production of hormones: Insulin is made by the pancreas, thyroxin by the thyroid, etc. Plant hormones, on the other hand, are synthesized in the cells of general organs—stems, leaves, roots, and flowers. Five different hormones have been identified in plants, with others, undoubtedly, awaiting discovery.

For as long as gardeners and botanists have been engaged in their affairs, they have observed the varied growth patterns of plants under different conditions of illumination. In full sun, stems are short and thick, the leaves closely spaced; whereas shaded stems become elongated, with poorly developed leaves. Seedlings grown in complete darkness bear little resemblance to those reared in light. The tall, thin, colorless stems of dark-grown plants support pale, undeveloped leaves—symptoms of a physiological condition called *etiolation.* And in most species, when light strikes one side of the stems, they bend toward the source of illumination, thereby re-aligning leaves to capture the energy.

The first plant hormone to be discovered was the substance causing stems to grow toward light—a physiological process called *phototropism. Tropisms* (Greek: *tropos,* turn) are growth responses to external stimuli.

120

Grown in full sunlight, pea seedlings bear well developed leaves on green stems. Seedlings of the same age but grown in total darkness display the etiolated condition.

In stems illuminated from above, cells undergo equal rates of elongation, resulting in vertical growth. But when lit from one side, stems change direction because cells on the shaded side grow faster than those toward the light. Phototropism is a common response in sun-loving species. For example, when such plants are placed indoors, near a window, stem curvature takes place. In some species, leaf petioles may also be phototropic. Most shade-loving species display little or no phototropic responses, an important factor in their selection as house plants.

The hormone controlling phototropism is named *auxin,* after a Greek word meaning "to increase". The chemical name for natural auxin, produced by plants, is indole-3-acetic acid, or *IAA.* Several synthetic substances (naphthaleneacetic acid, or NAA; 2,4-dichlorophenoxyacetic acid, or 2,4-D, etc.), having auxin-like effects when applied to plant tissues, have been studied and are used commercially.

Auxin's principal function is to stimulate increases in cell length, especially near stem and root tips (page 23). IAA is produced in cells of the stem's apical meristem and moves downward into the roots; as it does, its concentration decreases. In stems, the extent to which cells elongate is directly proportional to the prevailing concentration of the hormone.

When light strikes one side of a stem, auxin accumulates in the shaded side, causing the cells there to grow at the fastest rate. Thus, the plant growing indoors is forced to bend toward the window because light coming from that direction induces a re-distribution of auxin in the stem, resulting in uneven growth.

Although the underlying principle of phototropism seems uncomplicated, the details of how the process takes place are not understood. For example, plant physiologists are at a loss to explain what causes IAA to move from brightly lit to shaded cells. And the complex biochemistry of cell growth, including cell wall expansion, is still under investigation.

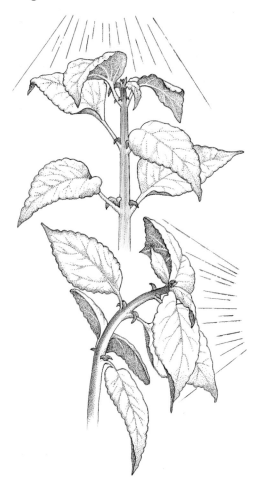

Phototropism. When plants are illuminated from above, auxin (represented by color) is evenly distributed across the stem, resulting in straight growth. Light striking one side of the stem causes auxin to migrate to the shaded side, resulting in more rapid cell growth than in the lighted side.

Growth of stem internodes (page 39) is promoted by another hormone, named *gibberellin* after the fungus *Gibberella* in which it was first discovered. The action of gibberellin on internode cells is also related to light intensity. In full sun, the hormone's effect on growth is somewhat restrained. Thus, while gibberellin promotes sufficient internode elongation to space the leaves, the structural stability of a squat growth form is maintained.

In low light intensities, however, gibberellin becomes more active, causing internodes to stretch. By so doing, the upper leaves are elevated to a position where they are better able to locate light, especially in competitive situations with surrounding plants. Interestingly, shade-loving species, being fully adapted to their preferred habitats, show no such reactions to low light intensities.

It appears to be a state of desperation that makes dark-grown, etiolated plants direct all of their energies into internode elongation as they search for a vestige of light. Gibberellin's unbridled stimulation brings about such growth. But rarely does the response result in a successful outcome since reserve food supplies are soon exhausted.

Responses to Gravity and Touch

When roots and shoots grow in opposite directions, they reflect contrary responses to earth's gravitational field. *Geotropism* (Greek: *ge*, earth), or *gravitropism*, are the names given the physiological process. Most roots are *positively geotropic;* that is, they grow in the direction of gravity. Stems, for the most part, are *negatively geotropic*, growing opposite to the gravitational force. When rhizomes, stolons, and some roots grow horizontally, they display *diageotropism* (Greek: *dia-*, across). And branches from roots and stems, developing at an angle from the vertical plant axis, are *plagiotropic* (Greek: *plagios*, oblique).

When a stem is placed on its side (perhaps a potted plant is accidentally tipped over), the apex soon returns to its normal direction of growth. It is known that, under the influence of gravity, auxin collects in the lower side of the stem, where it stimulates the cells to grow more rapidly than those across the top. But the mechanism of hormone migration to the underside of the stem is a mystery, since the weight of auxin molecules is less than the smallest objects that can usually move under the influence of gravity.

Although cells in stems elongate when high concentrations of auxin are provided, those in roots respond to very small amounts of the hormone. In roots placed in a horizontal position, auxin accumulates in the lower side, as it does in stems. However, it is the lesser amount of hormone in the root's upper cells that promotes elonga-

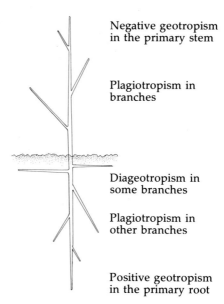

Negative geotropism
in the primary stem

Plagiotropism in
branches

Diageotropism in
some branches

Plagiotropism in
other branches

Positive geotropism
in the primary root

Geotropic responses in a plant axis and its branches.

Geotropism (gravitropism) in stems. Under the influence of gravity, auxin (color) accumulates in the lower side of a horizontally-placed stem. The high concentration of the hormone in the lower cells causes faster growth than in cells across the top.

tion and causes the root tip to grow downward. Recent studies have suggested that another hormone, produced in the root cap (page 37), is also involved in the geotropism of roots.

Geotropism plays an indispensible role in seed germination. In the soil, the embryos of randomly scattered seeds point in many directions; but soon after the root and shoot tips emerge from the seeds, their orientations are recognized and the appropriate geotropic response sets them on the right course. Imagine how impossible it would be to gardeners if every seed had to be planted in a "correct" position for them to germinate successfully.

Geotropic propensities are inherited, but how do roots and shoots acquire different responses to auxin concentration when, in each plant, both develop from the same fertilized egg? What causes the primary stem of a plant to grow vertically while its branches, responding in various ways to gravity, point in many directions? And what hormonal changes occur when branches growing from rhizomes change from a horizontal to a vertical position? Although the answers to such questions have not been found, there is no doubt that only through the various responses to gravity are different parts of a plant able to occupy the three-dimensions of their growing space.

Thigmotropism, the response to touch (Greek: *thigma*, touch), is seen in the ability of tendrils to grasp supportive objects (page 103). Coiling of a tendril results from faster growth of cells on the outside, away from the support, compared with those making the contact. Uneven distribution of auxin between the two sides of a tendril is presumed to bring about the growth differences, but how the slightest pressure can cause the hormone to migrate to one side is yet another unsolved puzzle concerning tropisms.

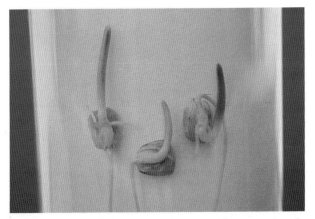

Regardless of a seed's position during germination, the emerging shoot quickly orients itself to grow in an upward direction. The principal root grows directly downward but smaller, branch roots show less of a positive response. Plant hormones control such growth patterns.

Other Growth Movements in Plants

Time-lapse photography compresses hours or days into seconds or minutes of viewing time. Through this technique, we can appreciate heretofore unseen marvels of plant life: The germination of seeds, flowers opening and closing, and the dance-like rhythm performed by growing plants.

By speeding the action, stem tips are seen to wave from side to side, or move in a spiral manner, rather than grow in a straight line. Such *nastic movements* (Greek: *nastos,* pressed close) result from cell growth at changing positions across the stem tip, pushing the stem in alternate directions. Nastic movements are controlled by hormones but are not direct responses to external stimuli, as are tropisms.

The tips of twiners (page 103) describe wide circles when they grow. Upon contact with rigid objects, they simply continue to spiral around them and, thereby, gain support. The opening and closing of flowers are also nastic movements. In the first instance, cell expansion occurs in the petal's upper surface; when the flower closes, cells in the lower petal surface increase in size. Recent studies with Venus' Flytrap (page 116) have shown that a similar system controls the movements of its leaf traps, the closure being an exceptionally rapid growth response. Each time a trap closes and opens, the leaf increases in size, since changes in cell dimensions are not reversible.

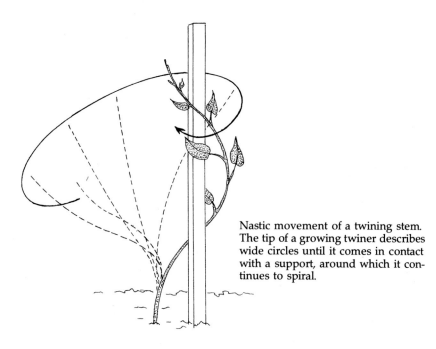

Nastic movement of a twining stem. The tip of a growing twiner describes wide circles until it comes in contact with a support, around which it continues to spiral.

Hormones and the Aging Process

When fruits ripen, they undergo aging, or *senescence* (page 83), a process also directed by hormones. When leaves are prepared for *abscission* (separation) from the stems, either individually or as a group in deciduous species, they also undergo the aging process. The words "abscission" and "abscise" come from the Latin, meaning "to cut off".

Since senescence, once begun, is irreversible, plants must possess strict controls over the process to prevent the premature demise of tissues, organs, or the whole organism. One of the functions of auxin, gibberellin, and another hormone, named *cytokinin*, is to inhibit senescence by maintaining both the functions and structural integrity of cells. Cytokinin (Greek: *kytos,* container—a cell; *kinesis,* movement) is principally responsible for promoting cell divisions. Acting antagonistically against those hormones, two other growth regulators promote the aging process. External environmental cues, such as seasonal changes in temperature or day length, and internal (biochemical) signals, shift the balance between the two sets of hormones.

The senescence-promoting substances are *ethylene,* a gas, and *abscisic acid,* so named because, originally, it was thought to promote leaf abscission in all species; later, it became clear that, most often, ethylene controlled the separation process.

When fruits ripen, various biochemical events take place. Color changes result from the breakdown of green chlorophyll and an increase in yellow, orange and red pigments (see Chapter 8). Tannins (page 90), which protect immature fruits against predators, give way to sugars—making some fruits attractive to seed-dispersing animals (page 26). And, primarily as a result of ethylene's influence, drastic changes occur in cell structure, including the breakdown of membranes and the softening of cell walls. The outcome of such processes is seen in the rapid deterioration of a fruit as it becomes "overripe", ready to release its seeds.

Unprotected by tannins, well-stocked with carbohydrates, and the cuticle layer degraded, the ripe fruit also becomes a perfect medium on which fungi can grow. Their presence eventually leads to the fruit's decomposition.

Ethylene is used commercially to ripen fruits, such as bananas, that are picked "green" for shipment. They are treated with the gas prior to being sold in retail stores. At home, some unripe fruits can be hastened to an edible condition by placing them in a paper bag with coarsely-chopped apple pieces. Injury of the apple tissues, as with damage to most plant organs, causes the cells to liberate ethylene. The bag should be closed to trap the gas, and kept at room temperature for

Six green persimmon fruits were picked the same day. At the end of one week, three that were placed in a bag with apple peelings—a source of ethylene—had ripened; the others, stored on an open shelf, had only begun to change color.

several days to obtain results. Not all fruit species respond to such treatment, but it is worth trying in every case.

Leaf senescence, prior to abscission, includes the breakdown of chlorophyll and weakening of cell walls at the base of the petiole, in a narrow band of cells called the *abscission zone.* In spring and summer, auxin produced in the leaf keeps the abscission zone intact. But low night temperatures and short days in autumn cue the leaves to reduce auxin production and increase the liberation of ethylene. The latter stimulates an enzymatic breakdown of cellulose walls and pectin in the middle lamellae—the "glue" holding cells together (page 20). The restriction of ethylene's destructive effects, only to cells in the abscission zone, illustrates the precise control plants exercise over their hormone systems.

In evergreen species, sequential abscission in the oldest leaves occurs by way of the same series of biochemical events. In a like manner, abscission of fruits and flowers takes place at appropriate times. During the development of some species' fruits, apple being an example, three periods when abscission can occur have been identified: "Post-blossom drop," "June drop," when fruits are still imma-

A microscopic view of a sectioned axillary bud and the base of its adjacent leaf petiole. The leaf's abscission zone appears as a dark band across the petiole.

ture, and "Pre-harvest drop," when many mature fruits fall to the ground. The application of synthetic auxin sprays increases yields by preventing early fruit abscission.

Control of Branching and Adventitious Root Formation

Gardeners know well that, to make their plants "fill out" with branches, stem tips must be removed periodically. As long as apical buds are present, they suppress the growth of axillary buds, especially toward the top of the stems. This physiological process is called *apical dominance.* Axillary buds toward the base of an intact stem are distanced from the effects of apical dominance and develop into branches.

Auxin, produced in the stem's apical meristem, exerts the inhibitory effect on axillary bud growth—a fact demonstrated when the apical bud is replaced with an artificial source of the hormone that is equally effective in bud inhibition. Trimming a plant simply eliminates the auxin source. In the case of basal buds, the natural decrease in auxin's concentration down the length of a stem is sufficient to reduce its suppressive effect.

Horticulturists are also familiar with differences between species' abilities to form *adventitious roots* (page 101) on stem or leaf cuttings. In some, root development is promoted by auxin naturally present in the cuttings. Other species must be treated with a rooting compound—a preparation of synthetic auxin. The method of vegetative propagation called *layering,* a modification of usual cutting techniques, involves the encouragement of adventitious roots on branches still attached to the parent plant. The advantage is that the stem piece, or "layer", is supplied with water and nutrients during the rooting process. When roots have formed, the layer is removed and grown independently. Again, auxin is involved in root formation.

An intact oleander stem (left) shows the normal pattern of apical dominance and suppressed axillary bud growth. With the stem tip removed, axillary buds freely develop into branches.

The success of propagation by stem and leaf cuttings depends on the species' ability to form adventitious roots. Shown here: African violet leaves.

Other Hormone Effects. Synthetic Growth Regulators

In the early stages of research on plant growth regulators, it was speculated that each physiological process is controlled by a different hormone. However, it soon became apparent that plants regulate numerous processes by employing only a few hormones, either individually or in combination with each other.

In addition to the roles played by the five hormones described above, the following can be added: Gibberellins control seed germination; ethylene promotes stem thickening, especially in seedlings; in some species at least, abscisic acid brings on dormancy, whereas gibberellin revives them from winter sleep. In some species, artificially applied gibberellin promotes flower formation and increases fruit size. Ethylene initiates flowers in pineapple and its relatives, the bromeliads (page 106) which, according to some reports, need only be enclosed in a bag with apple peelings for several days to produce flowers. Unfortunately, neither ethylene nor gibberellin are universally effective as "flowering hormones".

Plant physiologists have discovered several chemical compounds not produced by plants which regulate growth when introduced into certain plant species. Such substances are becoming increasingly useful in agriculture and horticulture. *Defoliants* promote leaf abscission and are favored by cotton growers, for example, since leafless plants facilitate the mechanical harvesting of cotton bolls. *Disbudders* cause flower buds to abscise and are sprayed on some species of ornamental trees and shrubs to rid them of developing, unwanted, troublesome fruits. One use of *growth retardants* is in the greenhouse floriculture industry for the production of dwarfed, potted plants such as chrysanthemum and poinsettia. Such chemicals inhibit the action of gibberellin on internode elongation, adding to the plants' aesthetic qualities and making them easier and cheaper to transport. *Herbicides* are most often used as weed killers. Some, such as 2,4-D, selectively destroy only broad-leaved species; others are non-selective in their activity. The search for plant growth regulators, first started as a subject of academic interest, has grown into a multi-million dollar business.

Environmental Control—Temperature

The controls hormones and other biochemical systems exercise over a plant's growth and development may, in turn, be regulated by seasonal environmental changes—especially changes in temperature and day length (photoperiod). This is the case in whole-plant leaf abscission (page 128).

For millennia, variations in day length with each passing month have been as regular as the sun's shifting position in the sky; and, in temperate zones, an annual period of winter's chill can be reasonably assured. The consistent occurrence of such events make them ideal conditions to which the physiology of plants is adapted, using them as external cues to turn on internal processes. Environmental cues frequently prepare plants for upcoming, adverse seasons in which dormancy is the only means of survival. But they also set in motion the first stages of reproductive cycles.

During winter dormancy, plant metabolism virtually comes to a standstill due, in part, to low temperatures that slow chemical activity. Botanists were surprised, therefore, to discover that several important physiological processes take place during winter and actually depend on the occurrence of reduced temperatures. For example, dormant winter buds must experience a cool period to prepare for eventual awakening in the warmer days of spring. The chilling process is believed to stimulate the synthesis of a hormone—possibly gibberellin—needed for subsequent growth. Exposure to days or months of temperatures below 50° F (10° C) may be necessary to overcome most bud dormancy. Apple, for example, requires 1,000 to 1,400 hours at about 45° F (7° C).

Such a requirement limits some species' geographic distribution to cool climates. However, selected varieties of temperate-zone fruits, such as peach and plum, needing only short periods of chilling, or none at all, are available for cultivation at latitudes where winters are warm.

It is important that a plant, deep in winter sleep, not awaken before spring, since new leaves and flowers are especially vulnerable to frost damage. To ensure the correct timing of dormancy's end, many species require a season of low temperatures followed by a period when day lengths are longer than those in winter. The second of these environmental cues is probably recognized by a subtle timing system within the bud scales (page 44).

In many such species, the springtime emergence of flowers precedes that of leaves—making the flowers more visible to pollinating insects and getting the reproductive process off to an early start. Incidentally, the flower buds are initiated before dormancy, in the latter part of the preceding year as a response to shortening day lengths.

A similar situation, in which reduced temperatures promote flower development, is found in many bulbous species (page 108). In autumn, tulip bulbs, for example, contain rudimentary flower buds. But for the flowers to complete their development by spring, the bulbs must first experience a drop in temperature, to about 50° F (10° C), for 13–14 weeks. A return to warmer temperatures promotes leaf and stem development and flower opening.

In temperate zones, the required temperature change occurs while the dormant organs are overwintering in the soil; but in warmer parts of the world, tulip bulbs must be dug up each fall and refrigerated to guarantee flowering. Hyacinth, daffodil and onion have similar requirements, although the precise low temperature and duration of treatment varies between species. Onions grown primarily for their bulbs may be prevented from flowering by maintaining the bulbs at warm temperatures throughout the year.

Vernalization

In some species, flowers are formed in spring as a result of the plant, or even the seed, being given a cold treatment, close to or at freezing temperatures, for several weeks. The physiological process is called *vernalization* (translated from Latin as "Belonging to spring").

Vernalization was first recognized in certain "winter varieties" of rye, wheat, and other cereal grains, the seeds of which must be planted in autumn and allowed to overwinter in the cold soil. Frequently, they germinate and spend the winter as small seedlings. Winter rye so treated, forms flowers within 7 weeks after seedling growth begins in spring; whereas it takes 14–18 weeks for flowering to occur in plants grown from unchilled seeds. Such a delay jeopardizes the plants' ability to complete the reproductive cycle before the end of summer.

Many biennial species require vernalization in order to produce flowers in the second of their 2-year life cycle. Examples include varieties of cabbage, kale, Brussels sprouts, carrots, celery, and foxglove. Most often, the vegetable crops are harvested at the end of their first year of vegetative growth; when cabbage, for example, is in the form of a tight "head" or leaf rosette. But, if left to overwinter, the plants make chemical preparations for the burst of stem growth, called *bolting*, leading to flower development by late spring.

The increased temperatures and lengthening days of spring promote the rapid internode elongation characterizing bolting, as well as the development of flowers at the top of the bolted stem. Such conditions have no effect on unvernalized plants, which remain in the rosette form and fail to flower. Gibberellin (the internode-elongating hormone) is probably produced during vernalization since unvernalized, but gibberellin-treated plants bolt.

Chrysanthemum is an example of a perennial species requiring vernalization. During winter and early spring, the young shoots, at the base of a chrysanthemum plant, respond to cool temperatures that must last for at least 3–4 weeks. The first, chemical stages of

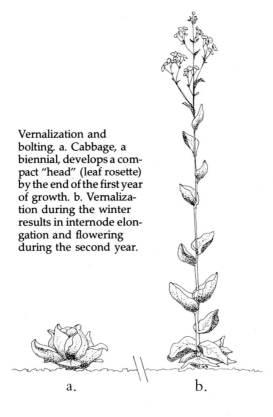

Vernalization and bolting. a. Cabbage, a biennial, develops a compact "head" (leaf rosette) by the end of the first year of growth. b. Vernalization during the winter results in internode elongation and flowering during the second year.

a. b.

flower development are begun at that time but become inactive until the shoots have grown. By late summer, when the plant is ready to complete the flowering process, the second phase is initiated by short day lengths.

The underlying, biochemical systems activated by low temperatures and preparing plants for flowering are beyond the present level of understanding. To complicate matters, chrysanthemum and many other species use such preparations merely to sensitize the plants to photoperiod, which, as we shall see, is an even more perplexing element in the flowering process.

Environmental Control—Photoperiod

Photoperiodism, the response of plants to changing lengths of day and night, was discovered about 1920 by W. W. Garner and H. A. Allard. Although several plant functions begin as a result of photoperiod, the initiation of flowers is the physiological process most frequently associated with this environmental control system.

When Garner and Allard recognized that plants are able to measure the passage of time, and thereby set in motion reproductive processes, they concluded that plants respond to the duration of daylight in each 24 hour period. The name they coined, *photoperiodism*, and other terminologies associated with the process, reflect such a belief. Unfortunately, later studies indicated that plants measure night lengths, rather than the daylight hours; but, for convenience, the "photo-" and "day" were left unchanged in the scientific vocabulary.

Angiosperms are roughly divided into three categories, based on their photoperiod requirements: Short-day plants, long-day plants, and day-neutral plants. *Day-neutral* species simply flower after some specific period of vegetative growth and are unaffected by day length. Examples include: Corn (Maize), cucumber, tomato, grape, pea, kidney bean, viburnum, and English holly. But consider the following and the *critical photoperiod* of each—the length of the daily light period necessary for flowering:

SHORT-DAY PLANTS—Flower when day lengths *are less than* the critical photoperiod of:

Chrysanthemum (*Dendranthema*)	15	hrs
Perilla (*Perilla*)	14	hrs
Poinsettia (*Euphorbia*)	12.5	hrs
Strawberry (*Fragaria*)	10	hrs
Violet (*Viola*)	11	hrs

LONG-DAY PLANTS—Flower when photoperiods *exceed* the critical photoperiod of:

Baby's breath (*Gypsophila*)	16	hrs
Dill (*Anethum*)	11	hrs
Red clover (*Trifolium*)	12	hrs
Sedum (*Sedum*)	13	hrs
Spinach (*Spinacia*)	13	hrs

Note that critical photoperiods vary between species. And "short-" and "long-days" are not defined by specific lengths of daylight hours, but by day lengths being less than, or more than the species' critical photoperiod. The responses of short- and long-day plants are summarized in the accompanying diagram.

Regardless of day/night lengths, a plant is only able to produce flowers when it has attained a *ripeness-to-flower*—a minimal vegetative size, necessary to support the weight of the blossoms and fruits, and a sufficient food reserve to supply the considerable demands of developing reproductive organs. Having reached that size, the plant is ready to be *photoinduced* by the correct day length.

In perennial species, ripeness-to-flower can be reached after one or more years of growth. But in annuals, seed germination and

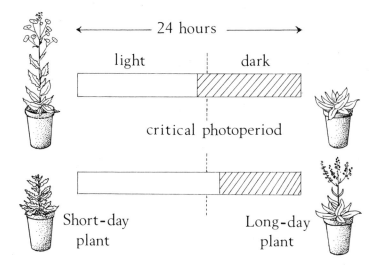

Photoperiodism. Short-day plants flower when day lengths are less than the critical photoperiod. Long-day plants flower when day lengths exceed the critical photoperiod.

development of vegetative organs must be completed in the early part of the year to prepare the plants for photoinduction when critical photoperiods are reached in the yearly cycle.

Most species require several successive days of photoinduction in order to change the activity of apical meristems from leaf to flower production—an astonishing switch, the mechanism of which continues to keep botanists guessing. In most species, once floral initiation has begun, the meristem never returns to making leaves.

Plant physiologists have determined that the biochemical system measuring the passing hours of day and night, and responding to the correct photoperiod, is located within the cells of mature leaves. When photoinduction occurs, such a system stimulates the synthesis of a "flowering hormone" that moves from the leaves to the apical meristems. At present, the timing process is imperfectly understood. Discovering the chemical nature of the flowering hormone ("florigen", as it has been named) continues to be one of the most sought-after goals of plant scientists.

The elusive florigen moves across grafts between photoinduced and uninduced plants, causing both to flower. It also moves between certain long-day and short-day plants; when either is photoinduced, the other also flowers. But, as with all plant hormones, florigen is effective in extremely low concentrations and is therefore difficult to detect. A prevailing theory among plant physiologists is that florigen is actually a special combination of gibberellin, auxin, and cytokinin, which makes its isolation even more difficult.

Whenever blossoms are desired, and since a flowering hormone to spray on plants is unavailable, floriculturists must be content with artificial photoinductions. Lights are used to extend natural day lengths; while heavy blinds are used to simulate short photoperiods. Several commercially-grown ornamental species, including chrysanthemum, are brought into bloom throughout the year by such methods.

Of all the physiological processes controlling plant growth and development, photoperiodism has the most far-reaching effects. Months after inductive day lengths initiate flower formation, reproduction ends with the dispersal of seeds. Since it takes time to develop flowers and, from them, fruits and seeds, floral initiation must be begun at an early date. In many species, it is imperative that reproduction be completed before the onset of winter, or summer's drought—conditions only seeds can survive. Thus one process, photoperiodism, connects two unrelated and separate circumstances—present day lengths and forthcoming adversity. Do plants not only measure the passing hours, but also anticipate the future?

Chapter 8

The Uptake and Use of
Water, Minerals, and Light

Osmosis, the Cell's Water Pump

Fashioned from earth, air and water, their life-sustaining processes kindled by the sun, plants build intricate bodies and manufacture foods to supply their every need. To such a self-sufficient life style, the name *autotrophic* ("self nourishment") has been given (page 32). Heterotrophic animals, fungi, and microorganisms, depend on products made by photosynthesizing plants, or resort to eating other heterotrophs to obtain such nutriments in second-hand form.

Autotrophs have few demands. Water and minerals that, in terrestrial species, are drawn from the soil, are added to carbon dioxide (CO_2) from the atmosphere. In the presence of light, photosynthesis, the single most important process on earth, is set in motion.

The journey that water makes from the soil to a plant's leaves begins when, by a process called *osmosis* (Greek: *osmos*, a push), it crosses into the root's epidermal cells (page 65). During osmosis, water molecules attempt to equalize their concentration on both sides of cell membranes when they move into or out of living protoplasm.

In most soils, small quantities of salts are dissolved in large volumes of water. Conversely, the protoplasm of epidermal cells contains lesser amounts of water in which salts, sugars, and other substances are concentrated. Thus, when water moves (diffuses) from the soil, where it is most abundant, it seeks to dilute the cells' solutions.

The system of equalization should also apply to salts, etc. that try to diffuse from the root's cells to the soil. However, cell membranes are selective in their permeability, permitting free inward movement of water but denying passage outward to most dissolved substances. It is such a preferential diffusion of water across membranes that makes osmosis work.

Water entering a cell is stored in the large, central vacuole (page 20), which expands and presses the cytoplasm against the rigid cell wall. When a cell becomes *turgid* (fully inflated), the rate of water

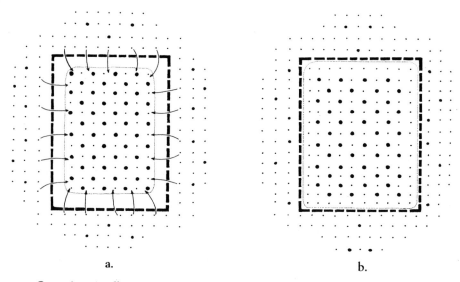

Osmosis. a. A cell is occupied by large amounts of dissolved substances (large dots), thereby reducing the space occupied by water molecules (small dots). Since water is more abundant in the dilute, external solution, it diffuses into the cell. The cell membrane prevents the loss of dissolved substances from the cell. b. Osmotic uptake of water creates turgor within the cell when the cell membrane presses against the cell wall. In these diagrams, the cytoplasm and vacuole are treated as one.

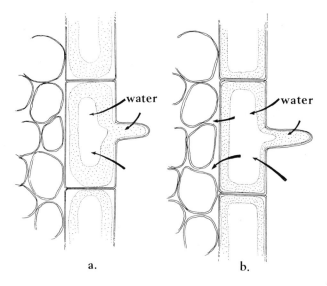

Water uptake by roots. a. Water enters a root's epidermal cells by the process of osmosis. b. As the vacuole becomes inflated, the cytoplasm is pressed against the cell wall creating an internal turgor pressure. At maximum turgor, water is squeezed out of the cell at the same rate as it continues to enter. The water moves into spaces between cortex cells.

uptake is slowed but does not come to a complete stop. Water continues to diffuse into the cell and simply displaces a comparable volume—the cell wall, counteracting internal *turgor pressures,* squeezes water out. Turgid cells are thus equipped with a "safety valve" which keeps them from inflating to the bursting point.

Cell turgor gives firmness to water-filled tissues. The difference between crisp and wilted lettuce leaves or celery petioles, illustrates the nature of turgid and non-turgid (flaccid) cells. Analogously, a bicycle tire has an expandable inner tube, inflated with air, that presses against the inelastic, surrounding sidewall. When deflated, the sidewall, like a cell wall, does not soften; it simply loses internal pressure.

Most plant species wilt in soils where significant quantities of salts have accumulated, even when adequate water is present. Such saline soils have a lower water content than root cells so the roots lose water as the direction of osmotic flow is reversed. In cells excreting large amounts of water, the vacuoles shrink and the cytoplasm is pulled from the cell walls—a condition called *plasmolysis* (Greek: *lysis,* loosening—of the cytoplasm).

Prolonged plasmolysis results in cell death. Yet the cells of seaweeds and Angiosperms adapted to coastal and desert salt flats are able to thrive in saline conditions, without suffering plasmolysis. Such an ability is attributable to such species' capacities to store salts at even higher concentrations than the external medium, thereby sustaining osmotic water uptake.

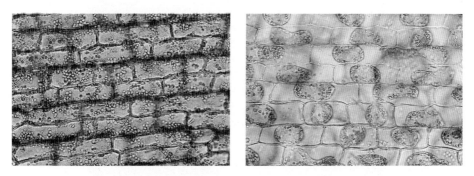

In turgid cells, chloroplasts are scattered throughout the cytoplasm that is pressed against the cell walls by an invisible, inflated vacuole. In the course of excessive water loss, the cytoplasm of plasmolyzed cells pulls away from the cell walls, forcing the chloroplasts into tight aggregations.

Development of Root Pressure

Near root tips, epidermal cells and their extensions, the root hairs (page 38), draw water from the soil by osmosis. When the epidermal cells are turgid, they discharge water into spaces between cortex cells—that being the line of least resistance for the escaping liquid. After the water works its way across the cortex, a second osmotic pump, the endodermis, directs it into the hollow, tubular cells of the xylem at the root's center.

Together, the epidermal and endodermal pumps push water across the root and up the xylem with a slight pressure, called *root pressure,* the effect of which is seen when liquid oozes from the cut stump of an herbaceous stem. Root pressure is also responsible for the droplets of water appearing, early in the morning, at leaf tips or on leaf margins. Such exudations are called water of *guttation* (Latin: *gutta,* drop) and emerge from special pores (hydathodes) evolved by certain species to rid themselves of excess dissolved salts.

Although root pressure may push water to the leaves of low-growing plants, it is insufficient to elevate water to hundreds of feet above the soil as in some trees. To accomplish such an engineering feat, a pulling force, generated in leaves, supplements the work done by roots.

Water droplets formed on leaf margins are the products of guttation—an exudation caused by root pressure.

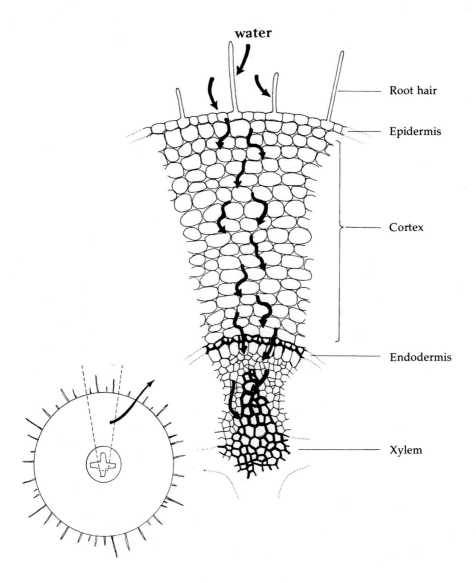

water

Root hair

Epidermis

Cortex

Endodermis

Xylem

The path of water across a root. The epidermis and endodermis function as osmotic pumps that move water from the soil, into the cortex and to the xylem at the center of the root.

Transpirational Pull

In leaves, mesophyll cells (page 67) with high concentrations of photosynthesized sugars, use osmosis to draw water from the xylem. When the water is later discharged from the turgid mesophyll, the sun's heat turns it to vapor saturating the leaves' internal spaces and ultimately escaping from open stomata by a process called *transpiration* (Latin: *trans*, across; *spiro*, to breathe).

Vaporized water, lost from the leaves, is replaced by liquid water pulled by the mesophyll cells from the veins. And since water molecules move in unbroken chains through the connected xylem of roots, stems and leaves, the drawing force of transpiration, called *transpirational pull*, is felt throughout the length of the plant.

To appreciate the combined effects of root pressure and transpirational pull, imagine a vertical tube in which water is both pumped under pressure at its base, and pulled by suction from above. The water would move at a considerable speed; but for the flow to continue, water must also be removed from the top of the system. In most plant species, about 98% of the water entering the roots is lost in the form of transpired water vapor from the leaves.

The magnitude of transpiration is impressive. A single, 48-ft. Silver Maple is estimated to transpire as much as 58 gallons (220 liters) per hour. A forest of temperate-zone, broad-leaved trees transpires about 8,000 gal. (30,000 l) of water per acre per day. An average-size, tomato plant transpires about 30 gal. (115 l) during its growing season; a corn plant, 55 gal. (210 l). Such quantities represent the plants' basic needs, to be supplied, via the soil, by rain and irrigation.

One can readily perceive the outpouring of transpiration when, on a hot, dry day, the air beneath a shady tree both feels cooler and is more comfortable because of its higher moisture content.

Although transpiration appears to be a wasteful process, plants cannot avoid losing water vapor through stomata that are opened to admit CO_2 for photosynthesis—a dilemma Nature has turned to advantage. Transpiration not only lifts vast quantities of water, against gravity, to the topmost leaves, it is also an effective means of transporting minerals from the soil to all parts of a plant in the xylem's transpiration stream. Furthermore, transpiration has a significant cooling effect on leaves exposed to full sunlight since water vapor escaping from a warmed, moist object, carries heat with it. The cooling sensation felt when perspiration evaporates from the skin, illustrates this principle.

Plants are far from being victims of uncontrolled water loss. They can quickly close stomata (page 69), especially when the quantity of water absorbed by roots falls short of that which is transpired. And, in some species, such features as matted epidermal hairs (page 86), the

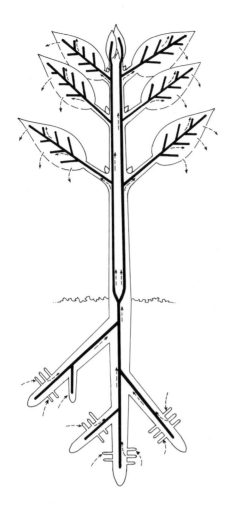

The path of water through the plant's inter-connecting xylem.

location of stomata in depressions in the leaf surface, and the enclosure of stomata within folded leaf blades, contribute to reducing transpiration.

The environment profoundly affects transpiration rates. Water loss is greatly increased by high daytime temperatures, a problem many desert species overcome by opening their stomata at night (and storing CO_2 in their tissues for use the following day). The difference in water content between the interior of leaves, saturated with water vapor, and the external air, is especially marked at times of low atmospheric humidities. Consequently, on dry days, transpiration increases and leaves quickly wilt. And, in a light breeze, air movement draws moisture from the stomata. However, when leaves are vigorously shaken, as by a strong wind (or even manually), the stomata close and transpiration stops.

Continuity of the water stream from the soil, through roots and shoots, and into the atmosphere is broken when a plant is uprooted, or stems are cut. Thus, a plant should be well watered immediately after transplanting to help it recover from the trauma of the move. Providing more-than-usual quantities of soil water compensates for transpirational losses occurring before the stomata have a chance to close, and ensures that damaged roots are plentifully supplied with moisture during their recovery.

The suction of transpirational pull places water under tension in the xylem of an actively transpiring plant. Thus, when stems are cut, air is drawn into the exposed vessels and blocks the flow of water. In the garden, flower stems should always be cut longer than desired, and re-cut to the correct length while holding the stems under water. If the stems are then quickly transferred to a vase, the transpiration stream continues uninterrupted, without the blossoms wilting. Scissor-type pruning shears or a sharp knife should be used to make clean cuts so the ends of the xylem vessels remain open.

Cold Hardening

In winter, when the leaves of deciduous species have fallen, water movement comes to a standstill. If the remaining water freezes in the cells, its expansion ruptures the delicate cell membranes—a condition from which there is no recovery. Plants prepare for winter with a process called *cold hardening,* part of which involves the accumulation, in the protoplasm, of sugars that function as "antifreeze" (page 86). In another phase of cold hardening, the permeability of cell membranes is changed, allowing water to leak into the intercellular spaces. In the event of ice crystals forming outside the cell walls, harm to the living protoplasm is averted.

Mineral Nutrient Needs

Gardeners are well acquainted with nitrates, sulfates, calcium, iron, and phosphates—some of the minerals plants require. And every gardener seems to have a list of favored fertilizers to satisfy their plants' needs. Whether such minerals are obtained from organic sources (composted vegetable matter, for example) or inorganic forms (commercial, crystalline or liquid blends of chemicals), the mineral elements are the same. The principal difference between the two sources is that decomposing organic matter slowly releases small quantities of unspecific minerals into the soil; whereas inorganic fertilizers are concentrations of selected elements, prepared for immediate and clearly defined uses.

Plant physiologists divide the required mineral elements into two groups: *Macronutrients* are those used in greatest quantities by plants; *micronutrients* are used in lesser amounts and, in some cases, are simply introduced as impurities in fertilizer mixes, or are dissolved in tap water.

MINERAL NUTRIENTS NEEDED BY PLANTS

Macronutrients	Chemical symbol	Available as:
Carbon	C	CO_2 (carbon dioxide) from air.
Hydrogen	H	H_2O (water).
Oxygen	O	O_2 (oxygen gas) and CO_2 from the air; and in some of the combined forms listed below.
Nitrogen	N	NO_3 (nitrate salts such as calcium nitrate) or NH_4 (ammonium salts such as ammonium sulfate).
Phosphorus	P	PO_4 (phosphate salts such as potassium phosphate).
Potassium	K	Potassium salts such as potassium phosphate.
Sulfur	S	SO_4 (sulfate salts such as magnesium sulfate).
Calcium	Ca	Calcium salts such as calcium nitrate.
Micronutrients		
Magnesium	Mg	Magnesium salts such as magnesium sulfate.
Iron	Fe	Iron (ferrous) salts such as ferrous chloride.
Copper	Cu	Copper salts such as copper sulfate.
Zinc	Zn	Zinc salts such as zinc sulfate.
Manganese	Mn	Manganese salts such as manganous chloride.
Molybdenum	Mo	Molybdenum salts such as potassium molybdate.
Boron	B	Borates such as potassium borate.

In addition, some species require traces of: chlorine (Cl), aluminum (Al), sodium (Na), silicon (Si), or cobalt (Co).

The elements used in greatest quantities by growing plants—carbon, hydrogen, and oxygen—are derived primarily from air and water. In nature, all other nutrients have their origins in earth's rock materials. Erosion slowly releases the minerals from rocks into the soil and, thence, into our planet's waters.

Each element plays specific biochemical roles; those of the

micronutrients being difficult to pinpoint since some such "trace elements" function in minute quantities in plant tissues, as the alternative name implies.

Clues to mineral nutrient functions are obtained, in part, from observable and predictable symptoms resulting from deficiencies of the individual elements. For example, the characteristic deficiency symptom of magnesium and iron is *chlorosis*—yellowing of the leaves—due to curtailment of chlorophyll synthesis. Magnesium is an integral element in chlorophyll molecules; iron must be present during production of the pigment.

Nitrogen is incorporated into the structure of chlorophyll, as well as amino acids, the small molecular units from which large protein molecules are made. Proteins are used to construct cell membranes, chromosomes (page 23), and enzyme molecules (page 93); all of which are vital to growth. It is understandable, therefore, that nitrogen-deficient plants display chlorosis of the leaves as well as stunted development.

Reduced growth, resulting from poor cell development, results from phosphorus deficiencies. Among other uses, the element is part of fatty membrane components called phospholipids. And during the production of DNA (page 18), the substance in cell nuclei bearing the genetic code, considerable numbers of phosphorus atoms are incorporated into the complex molecular structure.

The robust appearance of a tomato plant grown in a solution containing all the required mineral nutrients.

Magnesium deficiency results in yellowing (chlorosis) of the older leaves.

Chlorosis of young leaves is the first symptom of iron deficiency.

The poor growth of a nitrogen deficient plant includes: weak stems, undeveloped leaves, and reduced root development.

Phosphorus deficiency retards the growth of all plant parts—a consequence of imperfect cell development.

Calcium deficiency has the most drastic effect on growth. The element is needed to form pectin, the substance that bonds cell walls.

Calcium's role in the synthesis of pectin, the "glue" bonding cell walls (page 20), is vital to tissue formation in meristems. A shortage of calcium results in rapid death, or *necrosis* (Greek: *nekrosis*, deadness), of stem and root tips, and leaf margins.

The functions of potassium and metals such as copper, zinc, and manganese are too subtle to be reflected in observable deficiency symptoms. However, laboratory analysis of plant tissues has revealed that such elements are especially important as activators of key enzymes controlling metabolic pathways.

Since deficiency symptoms, especially those of micronutrients, vary between species and in response to changing environmental conditions, they are not always reliable indicators of specific plant problems. In addition, poor health may result from other factors, such as over-watering (especially of potted plants), the toxic effects of accumulated salts in the soil, high levels of air pollution, or invasion by pathogenic organisms. Or, in some circumstances, nutrients may be plentiful, but unavailable to roots because of prevailing soil conditions (see Soils, below).

Mineral nutrients return to the soil when leaves and branches periodically fall to the ground and decay, thereby completing one of Nature's most important cycles. Prior to leaf abscission, some of the nutrient elements (including nitrogen, potassium, and magnesium) are released from their bound form in protein, chlorophyll, and other molecules and transferred to the plant's growing tips for re-use. Since mineral relocation is in progress when older leaves turn yellow, gardeners can help their plants conserve nutrients by not removing discolored leaves for several days, until the minerals have been transferred.

During the course of growth, changes occur in a plant's nutrient needs—especially the relative quantities of nitrogen (N), phosphorus (P), and potassium (K). In the early stages of vegetative growth, greater amounts of nitrogen are needed to promote vigorous shoot development; whereas modest proportions of potassium are required for root formation. To encourage enlargement of the food-storage roots of "root crops", the ratio of these two elements may need to be changed in favor of potassium. At the time when a plant achieves a ripeness-to-flower (page 134), increased phosphorus and potassium, relative to nitrogen, promote the development of reproductive organs. Some species fail to flower when, at vegetative maturity, too much nitrogen is available; the plant's metabolic energies being directed into shoot and root growth.

Such relationships between nitrogen, phosphorus, and potassium, and the development of specific organs, underlies the reason for printing three numbers, called the *N-P-K ratio*, on packages of fertilizer. For lawn grasses and most house plants, recommended fertilizers have a proportionately high nitrogen content to promote leaf growth (20-5-5, being one example—20 parts nitrogen to 5 each of phosphorus and potassium). Whereas a product having a 0-10-10 ratio is a typical formulation designed for flower and fruit set. A fertilizer for root crops may have an N-P-K ratio of 2-12-10; and an all-purpose mix, 5-10-5.

Soils

The soil is one of the most important parts of a plant's environment, the medium in which roots are anchored and from which nutrients, water and oxygen are obtained. Soil is a complex mixture of inorganic materials, derived from the erosion of rock, and organic matter, called *humus*—the decomposed remains of plants and animals. The inorganic fraction is divided into three classes, defined by particle size:

Sand particles have a diameter between 0.02 and 2 mm.

Silt ranges from 0.002 to 0.02 mm.

Clay particles are smaller than 0.002 mm.

(A millimeter (mm) is 0.04 of an inch.)

Mixtures of sand, silt and clay are called *loams;* a sandy-loam, for example, contains proportionately more sand. In humus-loams, various proportions of organic matter are mixed with the other components.

The proportions of each material determine the water-holding capacity of a particular soil. *Water-holding capacity* (or field capacity) is defined as the water content of a thoroughly wetted soil after surplus water has drained off by gravity. Sandy soils retain little water; whereas the addition of humus increases the water-holding capacity, the moisture being held in tiny spaces (capillary spaces) within and between the organic particles. Capillary water is the principal source of moisture for roots.

A special problem is posed by soils containing large proportions of clay, the particles of which bear electrical charges that attract water molecules. The bond between water and clay, comparable to the attraction between the opposite poles of two magnets, is a difficult union for roots to break. Consequently, much of the water in a clay soil is unavailable to plants.

Another problem with clay soils is their lack of porosity. Dense packing of the fine particles leaves little space through which gases can be exchanged between the soil and the atmosphere; yet it is essential that CO_2 and other gases escape from below ground and that oxygen penetrates to a plant's roots. Water-logging, the saturation of spaces between soil particles thereby excluding air, is common

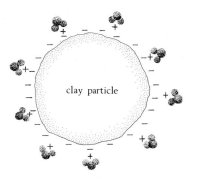

The surface of a clay particle bears negative electrical charges to which water is drawn. One side of each water molecule is positively charged. The bond between clay and water is difficult for roots to break.

in "heavy", sticky clay. In contrast, sandy- and humus-loam soils have a loose, porous quality favoring drainage and the diffusion of gases.

Soils have complex chemical compositions that determine, in addition to mineral nutrient content, their relative acidity or alkalinity—the *pH*. The "pH scale" is a numerical series from 1 to 14. A neutral condition is arbitrarily assigned a pH value of 7. Acidity decreases from pH 1 to 7; alkalinity increases from pH 7 to 14.

Most horticultural species grow favorably in soils at or close to neutrality. However, ferns, azalea, rhododendron, and camellia, for example, require acidic pHs between 4.5 and 5.5; whereas asparagus, spinach, cacti and other succulent species prefer mildly alkaline soils, to about 7.5 on the scale. Hydrangea tolerates a wide pH range, but its flower colors indicate the soil pH—the flowers become blue in acidic soils, pink in alkaline.

Most often, soils are made more acid by the addition of sulfur or organic materials such as peat moss or sawdust; limestone (calcium carbonate) is widely used to increase the alkalinity. The mineral content of irrigation water may alter the soil pH, particularly when the water evaporates and residual chemicals accumulate in the soil—a frequent, acute problem in arid regions.

Normally, rain leaches excess minerals deep into the soil and helps to restore the pH to more favorable conditions. But in some areas in recent times, industrial emissions are contaminating rain and affecting soil chemistry. When the air pollutant, sulfur dioxide, combines with atmospheric moisture, it falls to earth as a mild sulfuric acid solution, called "acid rain", that has a devastating effect on species ranging from simple nitrogen-fixing bacteria (page 114) to forest trees.

Increased soil acidity, whether from acid rain or natural sources, causes aluminum, manganese and iron to be liberated from harmless, insoluble forms in the soil to concentrations that slowly poison cells. Furthermore, free aluminum and iron cause phosphates to precipitate and interfere with the uptake of calcium by roots, thereby adding the deficiency of those macronutrients to other problems of plants in acidic soils.

At the other end of the pH scale, toxic amounts of molybdenum are released in alkaline conditions; both phosphates and calcium are rendered unavailable to roots when they combine to form insoluble calcium phosphate; and iron and manganese become tightly bound into chemical complexes. Since the resultant iron deficiency is fatal to many species, gardeners who work with alkaline soils must supply the metal in chelated form. *Chelates* (soluble organic compounds to which iron is bound) make the metal available to plants without toxic effects. The chelate is eventually broken down by microorganisms. Two commonly-used chelating agents are known by the acronyms: EDTA and EDDHA.

The Photosynthetic Apparatus

The electron microscope is one of the great inventions of the 20th century. The instrument makes possible tantalizing images of the intricacies of cell structure, at magnifications thousands of times their actual size. Chloroplasts, the site of photosynthesis, and other cellular organelles (page 19) possess such refined details that it stretches the imagination to fathom how each is made.

The photograph on page 152 shows two chloroplasts, separated by the walls of the adjacent cells in which they are located. Membranes enclose the organelles and form the parallel lines traversing the chloroplasts' length. Several stacks of short membranes are seen in each chloroplast. These are called *grana* (singular, granum) and are the exact location of chlorophyll and the other pigments that capture light—the primary energy source of plants and, ultimately, of all living things.

The leaves of higher plants contain various types of photosynthetic pigments. *Chlorophylls* occur in two forms—called chlorophyll *a* and *b*—both of which are green. *Carotene* is an orange-yellow pigment that is also abundant in carrot roots; and several *xanthophylls* range from shades of yellow to almost colorless, depending on their molecular structure.

When leaves "turn yellow", they simply lose chlorophyll which had previously masked the appearance of the orange and yellow carotene and xanthophylls. A purple-red pigment, called *anthocyanin,* may also be present in some species but does not participate in photosynthesis since it is stored in cell vacuoles. *Variegated* leaves are of horticultural interest for their inherited color patterns, formed by tissues in which these pigments occur separately or in unusual combinations.

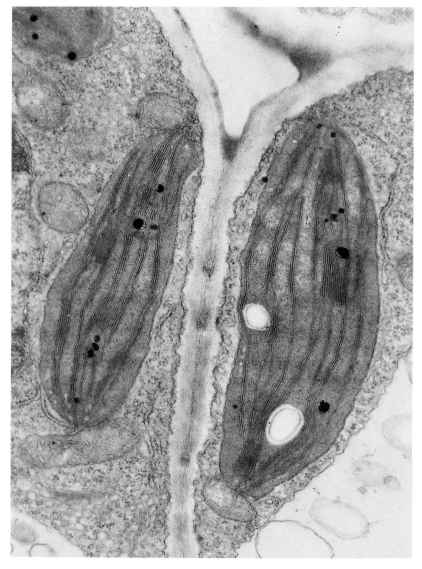

Magnified 36,800 times their actual size, two chloroplasts are separated by the walls of adjacent cells. Each chloroplast contains long, parallel, membranous structures connecting short membrane stacks, called grana. Chlorophyll and other pigments are concentrated in the grana. The right-hand chloroplast contains two starch bodies (light areas). The oval and elongated organelles, near the chloroplasts, are mitochondria—sites of cellular respiration. At bottom right, a portion of the cell's vacuole.

In autumn-colored leaves, chlorophyll molecules break down, unmasking the yellow carotene and xanthophylls. Some leaves, such as those of Liquidambar shown here, turn red when anthocyanin pigments add the final touch to the tree's colorful spectacle.

The inherited color patterns of leaf variegation result from the various pigments occurring separately or in combinations in mesophyll cells. Shown here: striped inch plant (*Callisia* spp.).

A rainbow's spectacle reveals that sunlight is composed of several colors. Of these, red and blue are captured by chlorophyll; whereas carotene and xanthophylls intercept only the blue-green part of the visible spectrum. At wavelengths represented by these colors, the energy of light is transferred, via the pigments, into the synthesis of foods.

Artificial illumination is only effective if it provides the blue and red wavelengths absorbed by chloroplast pigments. Ideally, incandescent bulbs, which radiate abundant red, should be supplemented with selected fluorescent tubes radiating blue wavelengths. And to achieve photosynthetic yields comparable to those in natural conditions, several lights are needed to provide high intensities; but care must be taken to control the build-up of heat.

Light Transformed into the Energy in Food

Light, heat, and electricity are different forms of energy, none of which can be stored for use by living organisms. During photosynthesis, plants absorb light and channel its energy into the formation of chemical bonds uniting atoms into molecular structures (page 52). The large-scale conversion of abundant sunlight into compact, energy-rich food molecules is the unique function of earth's flora. Scientists are able to transform light into electricity, but attempts to mimic photosynthesis have been in vain.

Food molecules—carbohydrates (sugars and starch), fats, and proteins—contain many chemical bonds, each representing a small "package" of stored energy. When foods are used in the biochemistry of cells, the bonds are broken and energy is released: Energy to construct other molecules, such as cellulose and lignin (page 21) needed for growth. Energy to make chromosomes dance through cycle after cycle of mitosis (page 23); to transport food in the phloem; regulate membrane permeability, and power countless other functions.

The process of extracting energy from foods, shared by all living things, is called *cellular respiration* and occurs in the cell's mitochondria (page 20). In higher animals, after food materials are digested, the energy-rich molecules are carried in the bloodstream to the body's cells. Among many physiological functions, the energy derived from foods makes muscles move and transfers messages through the nervous system.

Autotrophic plants hold the key to life on earth; they alone are the intermediaries between the sun and all other creatures. The principal role is given the leaves' tiny chloroplasts as proof that Nature is willing to entrust its most weighty responsibility to some of its smallest creations.

The Photosynthetic Process

The harvesting of sunlight by chloroplast pigments leads to a series of events in which water and CO_2 are used to synthesize simple molecules that, in turn, are used to build substances of increased molecular complexity.

Photosynthesis takes place in two stages. In the "light reaction", chlorophyll b, carotene and xanthophylls absorb and channel light energy to chlorophyll a whose electrons (charged, sub-atomic particles) are boosted to a high energy potential. In such an "excited" (energized) state, chlorophyll's electrons are diverted into a system that extracts and stores their energy for later use in the synthesis of sugars, etc. Despite the pigment's electron loss, it is quickly prepared to repeat the process when replenished with a fresh supply of the charged particles—water being the donor. The outcome of this electron transfer is the splitting of water molecules into the component hydrogen (H) and oxygen (O) atoms; the oxygen, in gaseous form (O_2), escaping into the atmosphere through open stomata. Astonishingly, the sequence of events described above is completed in a fraction of a second.

During the second phase of photosynthesis (CO_2-fixation), CO_2 from the atmosphere unites with the sugar, ribulose diphosphate. The product is split into two equal parts, hydrogen from the light reaction is added, and the resulting molecules of PGAL (phosphoglyceraldehyde) are then used as small "building blocks" for the construction of more elaborate molecular forms. Although CO_2-fixation is slower than the light reaction, millions of PGAL molecules are synthesized within minutes after light enters a leaf's mesophyll tissues.

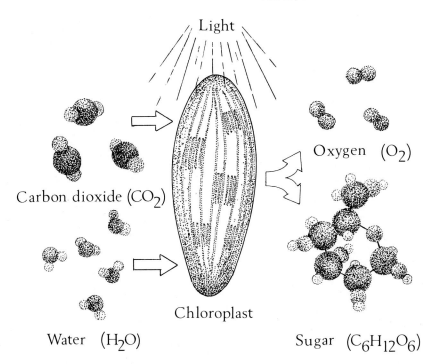

Light

Oxygen (O_2)

Carbon dioxide (CO_2)

Chloroplast

Water (H_2O)

Sugar $(C_6H_{12}O_6)$

Photosynthesis. Carbon dioxide and water molecules enter a chloroplast. Light splits water into its component hydrogen and oxygen atoms. The oxygen atoms are combined to form oxygen gas (O_2) that escapes into the atmosphere. The hydrogen and CO_2 are incorporated into molecules of sugar (glucose is shown here).

The first products of the metabolic construction process are several types of sugars, including ribulose diphosphate which the system regenerates, ready to repeat CO_2-fixation. Other sugars include glucose and fructose (both having the formula $C_6H_{12}O_6$ but differing in molecular structure); these may be combined to form sucrose ($C_{12}H_{22}O_{11}$)—common table sugar obtained from sugar cane and sugar beets.

Thousands of glucose molecules are united into long chains, forming the huge molecules of starch and cellulose. Although these two substances differ only in the manner in which the glucose units are joined together, they function in completely different ways. *Starch* is the principal food stored in plant cells, and can be broken into component glucose units when they are needed as an energy source in respiration, or for conversion into other plant products by the specialized activities of enzymes. *Cellulose,* on the other hand, once formed and incorporated into the structure of cell walls, is not normally decomposed for other purposes. Such multiple uses of glucose illustrate Nature's versatile but conservative mode of operation at molecular levels.

Following the production of sugars, the biochemistry of plants leads in many directions, some of which involve the introduction of mineral elements from the soil (nitrogen, sulfur, phosphorus, etc.) into the structure of certain molecules. Of the thousands of products thus formed, several that have been discussed in this and previous chapters are listed in the diagram on page 158.

An Outline of Photosynthesis

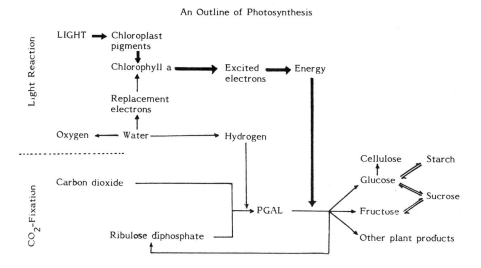

Gas Exchange with the Atmosphere

The liberation of oxygen during photosynthesis is as important to the well-being of living organisms as the conversion of light into the energy stored in chemical bonds. Cellular respiration, for example, occurs in most species[1] only in the presence of oxygen. The process removes oxygen from the atmosphere and releases carbon dioxide as a waste product. In photosynthesis, the converse is true. Thus, through the exchange of the same gases, the two processes serve to compliment each other.

On a global scale, oxygen is used for more than cellular respiration; all burning and rusting processes, for example, consume the gas. An essential role of photosynthesis, therefore, is to replenish the atmosphere's oxygen, a supply that would gradually diminish if autotrophs did not exist.

1. Among several organisms that carry on respiration in the absence of oxygen (anaerobic respiration), yeasts reduce sugars to ethyl alcohol (C_2H_5OH) and CO_2 by a process called *fermentation*. Yeast fermentation is employed in the production of alcoholic beverages; and when making bread, the released CO_2 is trapped in the bread dough, causing it to "rise".

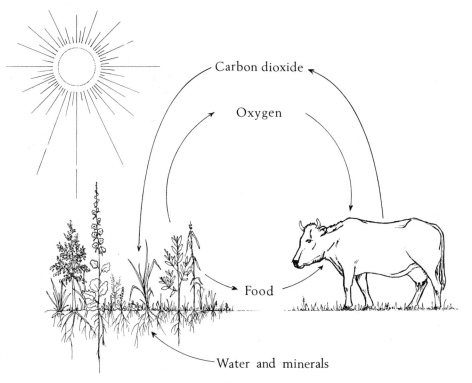

Carbon dioxide and oxygen are exchanged between plants and animals by the processes of photosynthesis and cellular respiration. Food, the principal product of photosynthesis, sustains all living organisms.

It has been hypothesized that, millions of years ago, photosynthesizing plants outnumbered other forms of life and were responsible for depleting the atmosphere of CO_2, to the present level of 0.03%. The gas was used to construct molecules forming the plant bodies. After the ancient flora died, the remains were converted into "fossil fuels"—coal, oil, and natural gas—through the work of geological processes. In those forms, vast amounts of CO_2 were locked in underground deposits for eons of time. That is, until mankind's widespread use of fossil fuels as an energy source began to undo the work Nature took so long to accomplish.

The burning of fossil fuels liberates the energy of sunlight, first captured by plants in prehistoric times. But in so doing, the ancient stores of CO_2 are also released. Scientists warn of a possible "greenhouse effect"—a rise in atmospheric temperatures due to the sun's heat being trapped below an increasingly dense layer of CO_2 and other air pollutants. As happened in the past, photosynthesis by abundant vegetation could remove the CO_2 from the atmosphere.

However, mankind's wholesale destruction of earth's great forests and other floras in recent times runs contrary to natural principles, thus adding to environmentalists' concerns.

The advent of photosynthesis opened the way for Nature to spawn the magnificent array of creatures inhabiting our living planet. Of these, evolution's most complex product, the human species, owes a special debt to this remarkable process. Photosynthesizing plants supply materials for food, clothing, and shelter; they filter the air; provide fuel for cooking and warmth; and, in fossilized form, yield the energy that powers marvels of another kind—the inventions of man's technological genius.

An Outline of Plant Metabolism

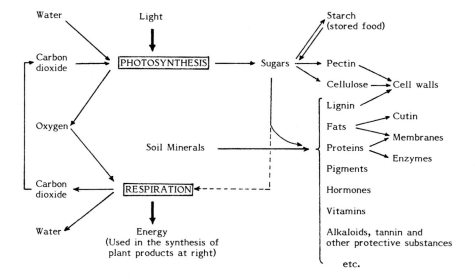

V. REPRODUCTION

Prologue

To gardeners, the botanist's definition of a flower—a shoot, modified for reproduction—may seem prosaic and hardly befitting the beauty of a hybrid rose or tropical orchid. But terse, factual, unsentimental descriptions are a part of scientific tradition. In truth, flowers are short branches bearing specially adapted leaves, and reproduction is the sole function for which flowers evolved; the pleasure they bring to mankind is coincidental.

Flowers are also clever lures, not the innocent beauties poets would have us believe. Casting camouflage aside, most flowers ostentatiously advertise their presence. Brightly colored petals and fanciful shapes flash in vivid contrast to subdued, leafy backgrounds, beguiling insects and other small animals into close floral inspections. Floral aromas fill the air—whether sweet scents attracting bees, or putrid odors to which carrion flies mistakenly flock. Convenient landing platforms, formed by petals, are provided for insects to rest from flight. And when the tiny animals are tempted to probe deeper into the flower's structure, they become unwitting assistants in the plant's reproductive process.

Not all flowers are so cunning. Several Angiosperm species, including grasses, bear inconspicuous blossoms that, like conifer cones, mushrooms, and the tiny capsules on mosses, disseminate their reproductive products by wind rather than animals. Growth of such modest structures, compared with showy but ephemeral flowers, takes less energy and food from the plant's reserves, and eliminates the need for complicated biochemical pathways to produce bright pigments and exotic perfumes.

Regardless of a reproductive structure's form or strategy, the function is the same. Reproduction is the means whereby life, the mysterious property of organisms, is transmitted from one generation to the next.

Offspring begin their lives from reproductive cells donated by parents—cells in which genetic instructions map lifetime growth potentials, developmental patterns, physiological activities, and the special adaptations characterizing the species. And, at maturity, the cycle is completed when an organism passes the bequest of its parents to progeny of its own.

Plants produce two types of reproductive cells. The first is the fine, dust-like particles issuing from brown spots on the underside of fern leaves, or from inside mushroom caps, for example, and are called *spores.* Under optimum conditions, each microscopic, one-celled spore grows into a many-celled plant by way of repeated mitotic divisions.

The second reproductive cell is formed during sexual reproduction, a more complicated system for producing offspring since it requires a population to be divided into male and female members or, at least, the development of distinct male and female structures on individual plants. The sex cells, or *gametes* (Greek: *gamein,* to marry), are incapable of growing directly into new plants, as do spores. In sexual reproduction, a *sperm* (male gamete) unites with an *egg* (female gamete) to form a single cell, called a *zygote* (Greek: *zygotos,* yoked, united), that then undergoes divisions leading to a multicellular organism.

The great problem inherent in sexual reproduction is the transfer of sperm to eggs of the same species. Since most primitive plants live submerged in water or in moist, terrestrial habitats, water provides a medium through which the sperm can swim. The egg is stationary, retained in the reproductive structure bearing it. The male gamete's search for a compatible egg is made easier when the latter excretes a chemical attractant that is believed to be specific for each species.

In many terrestrial habitats, water is not plentiful at the time of reproduction. Thus, to ensure the success of gamete transfer in Gymnosperms and Angiosperms, the sperm move through liquid-filled pollen tubes that first make contact with the awaiting egg. Reproductive processes, vital to the future of each species, have evolved to a state of extreme precision.

A few primitive species reproduce by spores alone; others use only sexual methods. However, during the life-cycle of most plants, gametes are formed in one phase; spores in another. Such is the case with the Angiosperms.

Spores are the principal units by which primitive species are propagated and dispersed to new habitats. During the evolution of Gymnosperms and Angiosperms, on the other hand, seeds replaced the spores as dispersal and propagative structures. In their dormant state, spores and seeds are equally suited to survive long periods of desiccation. But on germination, seeds have a head-start due to the presence of a rudimentary plant (the embryo) and storage tissues (page 27) which reduce the seedling's dependency on outside food and nutrient sources.

Mankind shares the transitory life common to all creatures. Whether favored with the longevity of the Bristlecone pines (page 61) or having a life-span of only a day, each organism's end is as certain as its beginning. Nature's great paradox is that such creatures, individually having a finite existence, should be chosen as vehicles to perpetuate life through eons of time. The apparent contradiction is reconciled by reproductive processes which range in complexity from simple cell divisions (as in unicellular organisms) to the elaborate methods of flowering plants, described in the following chapters.

Chapter 9

From Flowers to Fruit

Flower Parts and their Functions

One of the special rewards of day-to-day work in a garden is being witness to the unfolding reproductive cycles of plants. At first, the gardener may notice that some stem tips are "just a little different" than usual; or that axillary buds seem to have grown extra plump overnight. Such subtle changes are signs that reproduction is well underway; the biochemistry of flower development having been initiated some time ago, perhaps by photoperiod (page 133).

In subsequent days, recognizable flower buds take shape. Although still tightly enclosed by green scales, the buds swell with pent energy until, no longer able to contain themselves, they burst into full bloom. As layer upon layer of intricate structures unfurl in each opening flower, the scintillating, seductive display is more than adequate recompense for the gardener's patience and labor.

But such glory is short-lived. All too soon, the fragile flower parts wither and fall away. A central axis, that was least conspicuous in the blossom, now assumes the leading reproductive role as it slowly enlarges to become a fruit—a protective container in which seeds are nurtured.

The beauty and complexity of a flower takes shape within the confines of the bud. At the tip of the flower stalk, or *pedicel* (Latin for "little foot"), an apical meristem fashions the floral structures and arranges them in rings (whorls). The stem tip bearing the flower parts is called a *receptacle*. In a *complete flower* (having all of the customary parts), the outer whorl is called the *calyx* and consists of several *sepals*[1]. Most often, the sepals function as temporary, protective scales around the unopened bud. They may be shed by the opening flower, or curl backward as the next ring of modified leaves unfolds.

1. Most gardeners are familiar with the names of the flower parts, but not their word origins—*Calyx* and *sepal* are both derived from Greek words for "covering". *Corolla:* Latin, "little crown"; *petal:* Greek, "thin plate". *Stamen:* Latin, "thread", like the warp in an upright loom; *anther:* Greek, "flower"; thus, *perianth* literally means "around the flower". *Pistil:* Latin, "pestle"—

Anther + Filament = Stamen

Parts of a flower.

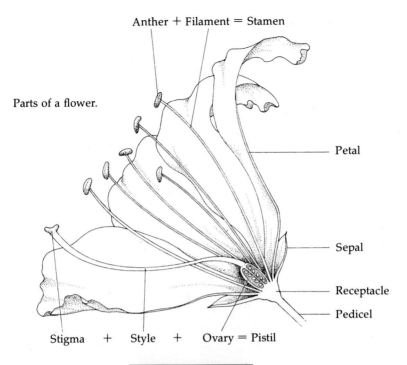

Petal

Sepal

Receptacle

Pedicel

Stigma   +   Style   +   Ovary = Pistil

Sepals, that form a protective cover over an unopened rosebud, fold back when the petals are ready to unfurl.

The *corolla* is composed of one or more layers of *petals,* and is the flower's show-piece—the visual attractant catching the attention of passing insects, birds, and people. Together, the calyx and corolla constitute the *perianth* of a flower. The petals may be white or brightly pigmented. In some species, the sepals may also possess colors other than green. Flower color frequently relates to the type of animal visitor the blossom receives—bees tend to be attracted to blue and

referring to similarities in the shape of the two objects; *stigma:* Greek, "spot"; *style:* Latin, "stylus"—a narrow, pointed instrument such as an etching needle. *Ovary:* Latin, "egg"; *ovule:* Latin, "little egg"—both of these are used broadly, referring to their general shapes.

violet, hummingbirds to red, but not exclusively so.

In the flowers of several species, the perianth consists of only one whorl of modified leaves called *tepals.* Tulips, for example, possess tepals that change from green to other colors during their development. Blossoms lacking one or more of the usual parts are called *incomplete flowers.*

The male reproductive structures are called *stamens,* each consisting of a stalk, or *filament,* which bears an *anther* at its tip. Within the anther, *pollen* develops. The fine, dust-like grains of pollen contain two cells, one of which eventually divides to form sperm cells. Each species of flowering plant may be identified by the distinctive shape of its pollen and the elaborate sculpturing of its outer wall, seen only with powerful microscopes. The pollen wall is so resistant to decay that fossil pollen, retrieved after being buried for thousands of years, have provided an accurate record of early-day floras throughout the world.

The colorful parts of tulip flowers are called tepals since no clear distinction between sepals and petals can be made.

A lily's reproductive parts—pollen-laden anthers and a pale green pistil—are clearly displayed against the pure, white petals.

The female part of a flower is the *pistil,* and is divided into three sections: At the top, a *stigma* or sticky receptive surface to which pollen adheres; an elongated *style* elevates the stigma into a favorable position for pollen collection; and, at the pistil's base, an *ovary.* The ovary ultimately becomes a fruit and contains one or more undeveloped seeds, or *ovules,* in each of which an egg waits to be fertilized by a sperm (see The Reproductive Process, below).

There are limitless variations in the shape and color of the flower parts, especially the petals. Sepals, petals, and tepals may occur as separate units, as in pansy and poppy flowers. Or they may be partly united into tubes (the corollas of foxglove and penstemon, for example), or the fanciful shapes of many orchid species. In daffodil and other *Narcissus* species, the corolla tube forms a prominent, crown-like outgrowth called a *corona*.

The perianth parts of many species (lily, crocus, and forget-me-not, for example) are symmetrically arranged around a central point in the flower, like the spokes of a wheel—an *actinomorphic* arrangement (Greek: *actinos*, ray). Snapdragon, sweet pea, and salvia, on the other hand, have flowers of irregular (or *zygomorphic*) forms. To botanists, flower shapes are important characteristics for the classification of Angiosperms into families, genera and species; as are the numbers of parts in each flower, their arrangements and sizes.

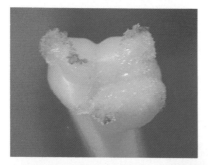

Pollen grains adhere to a lily's sticky stigma.

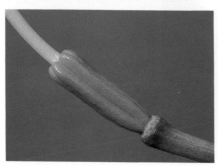

The ovary—the basal portion of a pistil, and future fruit.

A sectioned tulip ovary with immature seeds, or ovules.

Pollination by Animals

Having been attracted to a flower by its color, shape or aroma, an insect or other small animal may become an agent for *pollination* (pollen transfer). In some species, particular visitors are the only means by which pollination can be accomplished. Poised on flexible filaments, the anthers dust pollen on the animal's body. And as the little courier makes its rounds to other blossoms, some of its pollen load is brushed on their stigmas. However, for such a method to be successful, both the anthers and stigma must be strategically positioned to make contact with the animal as it probes the flower.

Differences between species' floral designs, including overall shape and the exact placement of stamens and pistils, are the products of natural selection. Many flowers are precisely engineered to match the body forms of the animals participating in pollen transfer. Because of their shape, some flowers are exclusively pollinated by a single species of bee, wasp, or fly—a precarious dependency since extinction of the insect would leave little hope for the plant species' survival.

Pollen transfer by animals is a more rapid, direct, and certain process than its random dispersal by wind or flowing water (page 172). Honey bees and hummingbirds, for example, move quickly between flowers, thereby dispersing pollen before the flowers wilt and stigmas become unreceptive. And, it would seem, even the smallest insects possess the desire to fly consistently between flowers of the same species—the most important requirement for successful pollination.

Poinsettia's small, yellow flowers would go unnoticed by animal pollinators if it were not for the brilliant red, modified leaves, called *bracts*, below them.

Road Maps and Rewards

For their assistance in plant reproduction, pollinating animals are rewarded with food—nutriments upon which the flower visitors become dependent. Thereby, one of Nature's most fascinating partnerships (or symbioses, page 90) is established. Some of the pollen may be eaten by insects such as beetles, or carried by honey bees to the hives to feed their young. But more significant, special glands (*nectaries*) at the base of the petals, pistil, and stamens exude droplets of a nutritious liquid, called *nectar*. It is during the pollinators' attempts to reach the nectaries that they inadvertently pick up and transfer pollen.

In some flowers, such as the tiny blossoms of Queen Anne's lace (*Daucus carota*) and other members of the carrot family (Umbelliferae, or Apiaceae), the nectar is clearly visible as a shiny drop in the bottom of the opened flowers. Such nectar is easily reached by ants and insects with short proboscises (flies, wasps, and beetles, for example). Other species typically conceal the nectar at the bottom of deep, cup-shaped or trumpet-like flowers. Or the food may be even more difficult to reach in the bottom of tubular, floral projections, called *spurs*—the distinctive features of columbine (*Aquilegia* spp) flowers. Such nectar is available only to insects having long proboscises (honey bees, bumblebees, butterflies, and moths) or to the extended tongues of hummingbirds.

Species having concealed nectar play games with their visitors. To feed from nectaries obscurely located within a flower, a pollinator is sometimes forced to assume a contorted position, to release a spring mechanism, or to move in such a direction that brushing against the reproductive parts is ensured. While enjoying a meal, bees may have to hang upside-down in some flowers—and, at the same time, have their bellies sprinkled with pollen. Insect visitors to such species as Mountain Laurel (*Kalmia latifolia*) are peppered with pollen when anthers on tensed, spring-like filaments, are released by the animal's touch. The elaborate floral structures of many orchids are so shaped that an insect, having been lured into a tiny opening, must work its way through a series of chambers or tubes in which pollen is scraped from its body and new loads are added. And while hummingbirds gracefully maneuver before pendant flowers, their feathers are

The long, tubular flowers of Torch Lily (*Kniphofia* spp.) are ideally suited to pollination by sunbirds in South Africa, their native habitat.

doused with sticky pollen from suspended, overhead anthers.

After an insect has settled on a landing platform formed by the lower petals of the flowers of many species, little is left to chance in aiding the pollinator to find its tasty reward and transfer pollen. The petal's color patterns, called *nectar guides,* are virtual road maps directing visiting insects to the sweet food, past waiting stamens and pistils. Converging stripes of contrasting colors, rows of dots, a brilliant circle of color at the flower's center, or star-like patterns provide visual clues that the insects instinctively follow. Some flower pigments reflect the sun's ultraviolet light—wavelengths visible to many insects, but not to the human eye. When nectar guides are painted with such pigments, the flower's pollinators are dazzled by marvelous, iridescent patterns to which we are not privileged.

No time need be wasted by visiting insects in flowers displaying nectar guides— contrasting color patterns such as those of iris and foxglove.

Inflorescences

Some flowers, such as tulip, form singly on upright stalks. Others occur in clusters called *inflorescences,* several types of which are described on these pages.

An inflorescence mistakenly regarded by many people as a single flower is the *composite head,* exemplified by daisy and its relatives in the Sunflower Family (Asteraceae, formerly Compositae). The typical composite head consists of a central cluster of small *disc flowers,* surrounded by a ring of *ray flowers* having conspicuous, strap-like petals. Frequently, the ray flowers are sterile, acting only as attractants to pollinators. In some Composite varieties, such as hybrid chrysanthemums and dahlias, there is no clear distinction between the two flower types in the inflorescence.

Branching patterns in some common inflorescences.

Spike. The flowers are attached to the main stem without stalks.

Acanthus spp.

Raceme. The flowers are attached to the main stem by short stalks.

Sweet broom, *Genista racemosa*

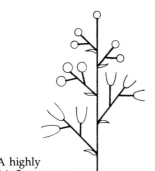

Panicle. A highly branched inflorescence.

Begonia spp.

Africa lily, *Agapanthus africanus*

Umbel. The flower stalks arise from one point at the tip of a stem.

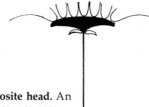

Composite head. An inflorescence composed of many tightly packed, small flowers.

Sunflower, *Helianthus annuus*

The shape of an inflorescence frequently relates to the pollinating animal's behavior—the approach and landing of a bee or butterfly, for example, or the hovering behavior of a nocturnal moth, hover fly, or hummingbird. Umbels and composite heads form ideal landing platforms for small insects, with plenty of flowers on which to feed. Whereas somewhat larger flowers, projecting from tall, solitary spikes and racemes, are favored by hummingbirds.

Inflorescences to which other species of birds are attracted have flowers bearing copious quantities of nectar and convenient perches that the birds can grasp while feeding. Bats forage for food at night, navigating with a unique sound system rather than by sight. Thus, the flowers that bats pollinate, and from which they feed, are clustered on isolated branches, away from the plant's leaf mass where the animal could become snared.

The individual disc flowers are clearly seen in a pyrethrum's inflorescence.

The calla lily (*Zantedeschia aethiopica*) is an inflorescence—a type of spike in which separate male and female flowers are borne on a central column called a *spadix*. The small male flowers are orange; the ovaries of the female flowers are pale green. The spadix is enclosed by a large, white, modified leaf (a bract) called a *spathe*.

Pollination by Wind and Water

Although wind-pollinated flowers are so inconspicuous they often go unrecognized in field and garden, they are no less wonderfully designed to fulfill the reproductive purpose. Most often, large numbers of the flowers are clustered on elevated branches where clouds of dry, powdery pollen can be wafted to receptive stigmas in distant inflorescences. In grasses, the anthers dangle on quivering filaments. And protruding from the tiny blossoms, disproportionately large, feather-like stigmas sweep the air as the flower stalks sway in each passing breeze.

Like delicate, white feathers, a grass' stigmas are prepared to capture wind-borne pollen.

In a wind-pollinated grass inflores-cence the individual flowers are small and inconspicuous. Brown anthers dangle from the flowers on fine fila-ments.

Of the vast amounts of pollen wind-pollinated species must pro-duce, only a small percentage lands in appropriate flowers. In view of the notorious unpredictability of wind's velocity and direction, it is difficult to understand how species can depend on this environ-mental factor to accomplish reproduction. But they do, and have successfully used the method for hundreds of thousands of years.

Despite the advantages of animal pollination (page 166), wind is the preferred mode of pollen transport for many species of Monocots—the most advanced group of flowering plants—including members of the Gramineae (or Poaceae, the grass family); Cyperaceae (reeds and sedges); and Juncaceae to which the wild rushes belong. Wind-pollinated Dicots are represented by such trees as beech (*Fagus* spp), oak (*Quercus* spp), birch (*Betula* spp), and elm (*Ulmus* spp) all of which form small, unnoteworthy flowers— sometimes in cone-like catkins, as do birch and hazel (*Corylus* spp).

Understandably, it is wind that transports the pollen of the most ancient pollen-producers—the Gymnosperms, from whose clustered male cones the grains are scattered like yellow dust. The pollination system of these plants evolved before the advent of insects, and has not changed since.

The need to produce large amounts of pollen, much of which is wasted, is also the fate of aquatic species of flowering plants whose pollen is disseminated by water. One of the most interesting examples is ribbon weed (*Vallisneria spiralis*), a popular aquarium plant, that forms female flowers on long stalks reaching to the water's surface. The male flowers are released from underwater inflorescences and float independently on the water like tiny sail boats. Propelled by gentle winds, some eventually make contact with the female flowers and transfer pollen.

The eel-grasses (*Zostera* spp and *Phyllospadix* spp) are marine Angiosperms that grow attached to rocks, along the shorelines of many continents. The unusual, thread-like pollen of these plants are adapted to tangle around the stigmas of the female flowers when swept there by waves.

Pollination Alternatives

Sexual reproduction can only occur between plants of the same species; foreign pollen landing on a stigma is incapable of delivering sperm to the eggs. It is believed that a flower obtains clues to the alien pollen's incompatibility from the grain's shape and chemical composition.

The far-reaching effects of sexual reproduction lie in the selective advantages of hybridization—genetic mixing occurring when gametes from different parents of the same species unite (page 77). In many species of flowering plants, elaborate methods have evolved favoring *cross-pollination* and consequent cross-breeding (outbreeding). These include: Self-incompatibility—chemical barriers in the stigma that treat a plant's own pollen as if it were from another species; spatial separation of the anthers and stigmas in a bisexual flower; or staggered timing of pollen release and the stigma's receptiveness in each flower.

Cross-pollination is also ensured when separate male (staminate) and female (pistillate) flowers are formed either on the same or different plants. When one plant bears both types of flowers, it is called a *monoecious* condition (Greek: *mono,* one; *oikos,* household). Monoecious species include: Corn (*Zea mays*), walnut (*Juglans* spp), filbert (*Corylus* spp), melons (*Cucumis* spp), and squash (*Cucurbita* spp). In *dioecious* ("two household") species, the two flower types are borne by separate individuals. Examples include: Willow (*Salix* spp), date palm (*Phoenix* spp), and pistachio (*Pistachia vera*).

So important is pollen transfer that some species possess self-pollinating, back-up systems for use when cross-pollination fails to occur, such as on cool days when insects are not active. In such cases, the flower's own pollen is not treated as a foreign strain. Although *self-pollination* precludes genetic diversity through hybridization, as a last resort it is preferable to no reproduction at all.

Male flowers of begonia bear a cluster of yellow stamens.

In begonia's female flowers, the pistil's yellow stigmas and styles are connected to an ovary, hidden below the perianth.

Some mechanisms of self-pollination include anthers that eventually sweep past the stigma as the stamen's filaments slowly curl, nasturtium (*Tropaeolum majus*) being an example. In other species, the filaments may elongate and carry the anthers past overhanging stigmas; or movement of the style may bring the stigma into contact with the anthers. In foxglove (*Digitalis* spp), the stamens are attached to the bell-shaped corolla. As the corolla is being shed at the end of the flowering period, the anthers may touch the pistil and thereby transfer pollen.

Although the mature flowers of many species open in a variety of environmental conditions, a few remain closed and undergo self-pollination in response to cold temperatures or certain photoperiods—a condition called *cleistogamy* (literally, a "closed marriage"). Such plants, to which insects are not attracted, conserve energy by not having to produce metabolically-expensive nectar during periods of poor growth. The flowers of some species of rock rose (*Cistus* spp) and salvia, for example, remain closed in cold climates; whereas several species of violet (*Viola* spp) develop closed flowers during the long days of summer; in spring, flowers are formed that open and are insect pollinated.

Many flowers remain closed at night and on rainy or overcast days, simply to protect the pollen against moisture. Self-pollination does not necessarily occur during those periods. On the other hand, flowers that do open after the sun has set are prepared for pollination by nocturnal animals. Such flowers have pale colors, visible in dim light; and most emit their strongest fragrances during the hours of darkness. On a midsummer's evening, in a garden where honeysuckle or night-blooming jasmines grow, one can share the ecstacy of nocturnal moths as, lured by sweet scents saturating the air, they are irresistibly drawn to the waiting blossoms.

The Reproductive Process

The stage is set for reproduction when, by one means or another, compatible pollen comes to rest on a flower's stigma. Of the two cells within a pollen grain, one is destined to grow into a long tube, a *pollen tube,* that penetrates the pistil's tissues in search of a microscopic opening in one of the ovules, located in the ovary. Germination and growth of the pollen tube is rapid and is promoted by food substances and hormones supplied by the stigma and style. Botanists still don't understand how a pollen tube locates the ovule's tiny pore; or the mysterious factor that diverts several tubes, growing from many pollen on the stigma, to separate ovules.

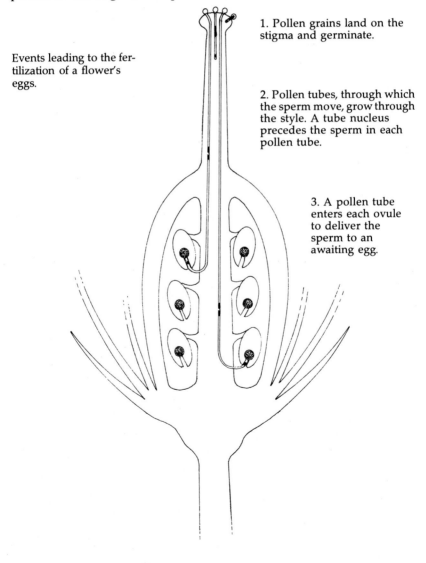

1. Pollen grains land on the stigma and germinate.

Events leading to the fertilization of a flower's eggs.

2. Pollen tubes, through which the sperm move, grow through the style. A tube nucleus precedes the sperm in each pollen tube.

3. A pollen tube enters each ovule to deliver the sperm to an awaiting egg.

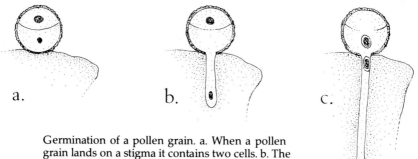

Germination of a pollen grain. a. When a pollen
grain lands on a stigma it contains two cells. b. The
nucleus of one cell controls the growth of the
pollen tube. c. The pollen's second cell divides to
form two sperm that move down the pollen tube.

The second of a pollen's cells divides to become two sperm that
move through the pollen tube and enter the ovule. Before the pollen
tube and sperm's arrival, each ovule must be equipped with an egg,
ready for immediate fertilization (see Chapter 10 for the method
involved). A *zygote* (page 160) is formed when one of the two sperms
unites with the egg.

The second sperm combines with another cell in the ovule. The
product of that union is a temporary food-storage tissue, called the
endosperm, used to nourish the zygote as it grows into an *embryo*—the
miniature plant within a seed (page 27). A portion of the endosperm
may persist in a seed and supply food to the growing seedling during
seed germination (Page 29).

As ovules grow and mature into seeds, they remain enclosed by
the ovary, which slowly enlarges to become a fruit—the *angeion* part
(Greek for "vessel" or "container") of the word Angiosperm, the
flowering plants' formal name. *Sperma* is Greek for "seed".

A well defined pistil sits at the heart
of each grapefruit flower. After fertilization of the eggs, most of the
flower parts drop off, leaving the
green ovary nestled in a tiny cup
formed by the sepals.

The development of most fruits and seeds occurs as a consequence of pollination and fertilization of the eggs. Embryo and seed growth apparently stimulate the production of hormones, including gibberellin (pages 123, 130), that promote fruit enlargement. When pollen tubes deliver sperm to the ovules in only one segment of the pistil, the resulting fruit merely enlarges on that side. Gardeners are familiar with such occasional, odd-shaped products.

Seedless Fruits and Unusual Embryos

A few species are capable of producing full-grown fruits without the stimulation of pollination and fertilization, and the development of seeds—a condition called *parthenocarpy* (Greek: *parthenos*, virgin; *karpos*, fruit). Examples of parthenocarpic fruits include the navel orange, common varieties of banana, oriental persimmon, and pineapple. Not all seedless fruits are parthenocarpic. Seedless grapes develop after pollination and fertilization, but the embryo soon aborts and the seeds fail to enlarge. Seedless plant varieties can only be propagated by vegetative means, including cuttings and grafts (page 60).

Normally, the embryo develops when the fertilized egg, or zygote, undergoes repeated cell divisions. However, in citrus and a few other species, certain cells in the ovule, other than the fertilized egg, may develop into a viable embryo. Thus, fertilization is bypassed and the resulting plant has a genetic composition identical to the single parent bearing the pistil and ovule. Such an unusual phenomenon is called *apomixis* (loosely translated from Greek: "a different mixing").

Fruit Types

As the ovary of some species develops into the fruit wall, or *pericarp*, it becomes increasingly fleshy and soft as the time for seed dispersal approaches. Or, in other species, the pericarp may dry and split open along clearly defined seams, as do pea pods when left to mature on their vines. Some dry fruits, such as poppy capsules, scatter their seeds through small openings, like a salt shaker. Others, including acorns and filbert nuts, become hard and remain closed; their seeds germinate only after the pericarps have partly rotted in the soil.

The outer layers of such fruits as peach, apricot, plum, and cherry become soft and delicious to eat, whereas the inner stone (or pit), which is also a part of the pericarp, is exceedingly hard and contains the fruit's single seed.

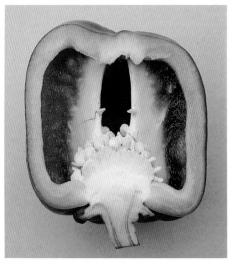

A pepper's fruit wall, or pericarp, is soft and fleshy at maturity. The seeds have a large cavity in which to develop.

Some fruits become dry and brittle at maturity. Poppy seeds are shed through holes at the top of the capsule.

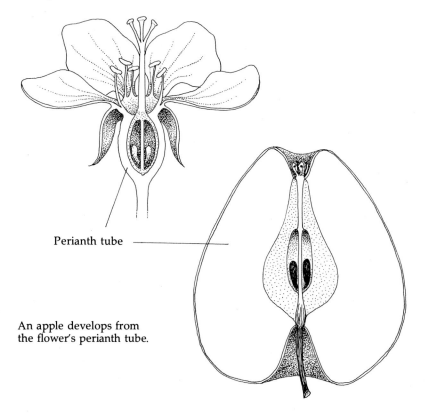

Perianth tube

An apple develops from the flower's perianth tube.

The edible part of most fleshy fruits is pericarp tissue—derived from the flower's ovary wall. But, in some cases, the tissues of petals and sepals, or an enlarged receptacle, become the special food material. In the flowers of apples and pears, for example, the basal portion of the perianth forms a tube fused to the sides of the ovary. When the fruits develop from such flowers, it is the perianth tubes (petals and sepals) that enlarge, rather than the ovary walls. Strawberry flowers, on the other hand, bear numerous pistils on a single receptacle. When the strawberry "fruit" matures, the receptacle grows to many times its original size, softens and becomes a bright red color. On its surface, the tiny, ripened ovaries (the true fruits) are borne; each contains a single seed.

Numerous pistils are borne on a strawberry's receptacle—a green dome at the flower's center. It is the receptacle that enlarges and ripens into a luscious "fruit".

To classify the various types of fruits, botanists have devised an elaborate system based on such features as pericarp structure, method of seed dispersal, and mode of fruit development. In such a system, the most basic criteria for classification are as follows:

A *simple fruit* is derived from a flower bearing a single ovary; examples include tomato, orange, grape, melon, and peach.

An *aggregate fruit* comes from a flower having many ovaries. Thus, an aggregate fruit is a collection of small fruits borne on a single receptacle; for example, blackberry, raspberry, and strawberry. The individual juice sacs in blackberries and raspberries are separate pericarps.

Multiple fruits are best exemplified by pineapple. The segments seen on the "fruit's" surface are actually separate pericarps, each derived from a flower with one ovary. Many such flowers crowd around the flowering stem. At maturity, the pericarps become fused into a single, edible mass.

The individual juice sacs of the boysenberry's aggregate fruit develop from separate ovaries, formed within a single flower.

Pineapple is a multiple fruit—derived from the fusion of several ovaries, each from a separate flower in the inflorescence.

Seed Dispersal

The transport of seeds away from the parent plant, hopefully to places where the seedlings can grow without competition for light, water, and soil nutrients, is the final act in the drama of reproduction. The fruit's function is to protect the seeds until they are ready for dispersal. The seeds may be shaken or vigorously ejected from the mature fruit when it dries and cracks open. Or they may be carried as a package, after the fruit has separated from the parent plant and rolls down a slope, or is carried by wind, water, or animals to some other location. The seeds are eventually released when the container rots or opens when dry.

The fruits of maple (*Acer* spp) and ash (*Fraxinus* spp) bear wings to help them fly with a spinning motion. Some small fruits, such as those of dandelion (*Taraxacum officinale*), receive widespread dispersal while suspended in midair on feathery parachutes. And many orchids produce large numbers of fine, dust-like seeds often blown great distances by wind.

Seeds and fruit scattered by water possess air-filled cavities and waterproof covers. The coconut's fibrous husk, for example, provides sufficient buoyancy to the heavy fruit and its single, large seed to enable them to float for many miles on ocean currents.

The presence of hooks and barbs is a reliable indication that a seed or fruit is waiting to hitch a ride on a passing animal. Either by snagging the fur or lodging in the animal's feet, the seed or fruit may travel some distance before it is shed or scratched off. Some seeds, such as those of the parasite mistletoe (page 112), are coated with a sticky substance that adheres to the feet and feathers of birds until the

Wind dispersal of seeds. Each dandelion fruit contains a small seed and bears a tiny parachute for dispersal to far-away places.

Seed dispersal by water. The coconut's buoyant fruit may travel great distances before being washed ashore and its seed germinates.

Animal dispersal of seeds. Attractive, red berries frequently contain small, hard seeds that pass, unharmed through birds' digestive tracts as a method of dispersal.

seeds are transferred to a tree's bark when the animal preens itself. In a similar fashion, water fowl may carry the seeds of marsh plants in mud stuck to their bodies. And the passage of seeds, unharmed, through animals' intestinal tracts may be an important factor in both seed germination and species' dispersal (page 26). In such a case, the food value the animal derives from digesting the fruit becomes an incentive to swallow and distribute more seeds.

The Cost of Reproduction

A significant portion of a plant's food reserves are used for the purpose of reproduction. When growing in field habitats, an average annual species spends 20–30% of its resources on flowering and fruiting; perennials use about half that amount. But the threat of death from nutrient or water shortages prompts many plants to increase food allocations (to 50% or more) to processes ensuring the perpetuation of the species.

The bigger the reproductive structures, relative to the plant's size, the greater the amount of food that must be accumulated during the preceding months or years of vegetative growth. Flowers, nectar, pollen, fruits, and seeds are each produced at an enormous energetic cost. Even among Gymnosperms, resources are taxed by the development of cones, prodigious quantities of pollen, and typically large numbers of seeds. But with the stamina of a pine, cedar, or redwood tree, production of an yearly seed crop does not cause irreparable harm.

Such is not the case with annual species whose death after seed formation is due, in part, to the stress the reproductive processes place on their metabolic systems. Likewise, some large, perennial Angiosperms end their lives after a single batch of seeds has been scattered—a good example being the Century-plant (*Agave* spp) that, after 6–15 years of growth, produces one, towering inflorescence, then dies. Such a sacrifice is indicative of the price living organisms are willing to pay for survival.

Plant Classification

Reproductive structures are among plants' most distinctive features used to classify species. Each plant group has identifying characteristics ranging from the gross morphology of flowers, to minute details of pollen and seed shapes, mode of embryo development, pollination systems, and floral pigmentations. In addition, leaf shapes, vein patterns, leaf and stem anatomy, the structure of epidermal hairs, chromosome numbers in cell nuclei, unique biochemical products, physiological traits, and habitat preferences may be added to the list of items making up a species' complete description.

Such lists continually grow with new information gathered from the close examination of specimens in herbaria and the field. Of special interest are the regularities and variations existing between plants of the presumed same species from different locations. When it is recognized that the disparities between populations are too great to correctly classify all specimens as members of one species, the group is split into two or more new species and re-named accordingly.

Occasionally, major differences between plants first thought to be closely related, may result in their reclassification into other genera or even different families. Such may be the case when heretofore unrecognized biochemical dissimilarities are discovered, or when new light is shed on the intricacies of developmental patterns, for example.

Taxonomists, the specialists who deal with the systematic classification of organisms, must continually keep abreast of research being conducted in other fields of study. As information on the thousands of recognizable plant groups continues to accumulate, taxonomic systems are forever being revised. Thus, the work of classification and nomenclature will never be completed, no more than the evolutionary processes creating species will ever cease to operate. Species are in constant flux, either developing in harmony with environmental change that, through geologic time, alters the course of natural selection (see Part III), or becoming extinct. In addition, the continual emergence of new gene recombinations, resulting from hybridizations, contribute significantly to plant diversification.

The scientific names of plants (genus, species) are certainly more reliable and universally accepted than common names that vary from country to country, and, more by whim than good judgement, may change between one century and the next. Even so, the very nature of taxonomy is such that the scientific names will always be subject to periodic review, and possible modification.

Many horticulturists find such changes somewhat disconcerting. But new names are given only after thoughtful study, and are meant to clarify our understanding of the evolutionary relationships existing between species and their close kin in other taxonomic groups. Shortcomings in the classification system simply remind us that while the Plant Kingdom's multitude of organisms has taken millions of years to develop, mankind's attempt to sort out the complex family ties has, by comparison, only just begun.

The "father of taxonomy" was the Swedish botanist, Carl Linnaeus (1707–1778) who devised the binomial system of nomenclature in which each plant is identified by a two-part name; for example, *Phaseolus vulgaris* L., the botanical name for the common bean. The first part is the name of the genus to which the plant belongs, and indicates its close relationship to other beans, such as *Phaseolus lunatus* L. (lima bean) or *P. coccineus* L. (scarlet runner bean). The *generic name* is a singular, Latinized noun and is always capitalized. The second part of the name, called the *specific epithet,* is often an adjective, and is written in lower case. Together, the generic name and specific epithet constitute the name of the species. The initial L., standing for Linnaeus, or another personal name printed in roman characters, is the name of the author of the plant name.

Unfortunately, when Linnaeus attempted to classify the bewildering array of plants into groups larger than genera, he chose only a single set of characteristics—the number and length of stamens. Such an artificial method of classification placed cacti and cherries in the same class, for example, and so was eventually replaced by a system designed to indicate the natural, evolutionary relationships among species. And it was Darwin's revolutionary ideas of the origin of species by natural selection that provided the theoretical framework on which taxonomists based their ideas.

In today's classification hierarchy, broad similarities between genera are used to group them into *families*. In turn, the natural affinities between certain families place them into one of several *orders;* orders are grouped into lesser numbers of *classes;* classes into *divisions* (or phyla). Thus, the system is like a pyramid, with few divisions at the top and several hundred thousand species forming the base. In accordance with the current International Code of Botanical Nomenclature, the preferred ending of all family names is -aceae (for example, Orchidaceae); that of orders is -ales (Rosales).

Species used in horticulture retain the name of the wild plants from which they were derived. However, cross-breeding, selection, cloning (page 72), and other procedures have resulted in the proliferation of cultivated varieties, or *cultivars,* of many such species. A true cultivar retains its distinguishing characteristics from generation to generation. The cultivar's name is either preceded by the abbreviation cv. (for example, *Papaver orientale* cv. Sultana) or placed in single, inverted quotation marks (*Hydrangea macrophylla* 'Europa').

The monumental task of discovering, sorting, describing and classifying earth's flora continues to be one of the major contributions botany makes to mankind's understanding of the world in which we live. More than being mere lists of species' names, taxonomic systems are useful scientific tools creating order out of chaos and, thereby, giving great satisfaction to the tidy human mind.

Chapter 10

Strategies of Inheritance

Genetics, the Science of Heredity

Family reunions offer excellent opportunities to see Nature's heredity systems at work: Gatherings of grandparents, parents and their children, cousins, aunts and uncles who are tied to a common ancestry by their *genes*—molecular units of inheritance, carried by chromosomes in the nuclei of cells. Distinctive hair and eye colors, facial features, and other traits that run in families, bear witness to the persistence of genes transmitted from generation to generation through the process of reproduction. So also do physiological propensities, some of which manifest themselves in the form of recurrent inherited diseases.

In plants, lineages can be traced by way of similarities in flower form, seed color, stature, hardiness in the face of cold or drought, unique biochemical products, and so forth. Each plant or animal characteristic is determined by a particular gene or set of genes. And when mixed by sexual processes, the new gene combinations result in offspring having recognizable family traits, together with unmistakable, individual qualities. This is no better illustrated than in the human species, where random assortments of several common features make each family member recognizably different from the others.

The underlying mechanism of gene mixing can be traced in plants more readily than in people, since cross-breeding can be controlled by selecting parents with specific traits; and several generations of plants can be produced within a comparatively short time, especially in annual species. It is not surprising, therefore, that the fundamental laws of genetics were first recognized by a keenly-observant gardener in the repeated hereditary patterns of his plants.

The precise work of Gregor Mendel (1822–1884), an Austrian monk, on the hybridization of common peas in his monastery's garden, and his innovative interpretation of the experimental results, has been recognized as one of the greatest intellectual accomplishments by a single individual in the history of science. Mendelian genetics revolutionized the biological sciences and provided theoretical support for Darwin's concept of speciation through natural selection.

The picture of how hereditary systems function was further enhanced when other workers discovered the detailed cellular processes involved in reproduction. From such knowledge, we are now able to appreciate the reproductive strategies plants use, and to follow the complex steps in the Angiosperms' and other plants' life-cycles.

Mitosis and Meiosis

The process of *mitosis* was described in Chapter 1 (page 24) as a cell replication process—the means whereby new cells are formed in meristems during plant growth (page 23). Each cell produced by mitosis is endowed with an exact copy of the parent cell's nucleus, including its chromosomes and genes.

Mitotic divisions are also responsible for the growth of unicellular spores into multicellular plants (page 160). However, when the time comes for a new generation of spores to be produced, a different cellular process, called *meiosis* (from the Greek, "to lessen"), is invoked. Like mitosis, meiosis is a divisional process occurring within cell nuclei. But where mitosis duplicates, and so precisely doubles the number of chromosomes, meiosis results in the reduction of each cell's chromosome number by exactly one-half.

A flowering plant's roots, stems, and leaves are composed of cells in which the chromosomes occur in pairs. The pairing takes place during sexual reproduction when one set of chromosomes is supplied by a sperm; the other by the egg. The same is true of our own body cells, each of which contain 46 chromosomes in matching sets of 23 from each parent. Plant and animal species are identified, in part, by their chromosome numbers. Among Angiosperms, chromosome totals range from 4 (2 pairs) per body cell in a species of *Haplopappus,* a type of daisy, to 264 in some grasses.

When chromosomes occur in pairs, the cell is said to have the *diploid* (twofold) number. A diploid cell dividing by mitosis forms daughter cells that are also diploid; that is, they are exact replicas. But when a diploid cell divides by meiosis, the products possess only half the number of chromosomes, called the *haploid* (single) number. A haploid cell may undergo mitosis to form haploid daughter cells, but can never divide by meiosis.

A fundamental difference between plants and animals is that in the latter, meiosis is reserved for the production of gametes (sex cells); whereas most plants use the process to form spores. To appre-

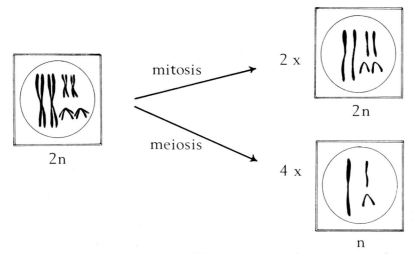

During division of a diploid (2n) cell by mitosis (top), the separate strands, or chromatids, of the paired chromosomes are shared by two daughter cells. Division by meiosis (below) results in four daughter cells, each receiving a haploid (n) set of chromatids from one member of each chromosome pair. The chromatids eventually duplicate themselves in the daughter cells to form double-stranded chromosomes.

ciate the roles of mitosis, meiosis, spores and gametes in plant reproduction, the life-cycle of a typical moss, fern, and flowering plant are considered in detail, below.

The Life-cycle of a Moss

Mosses congregate in shady, cool, damp places. Frequently, they form dense, green cushions of many thousands of individual plants that are only recognizable when a clump is gently pulled apart. In many ways, mosses are primitive organisms: They lack roots and vascular tissues (xylem and phloem); their leaves and stems have a simple anatomy, compared with higher plants; and because of an inability to form lignified, supportive tissues, they are low in stature. The little, green moss plants are composed of haploid cells, having grown from haploid spores by mitotic divisions.

In a group of mosses, about half the members are male and produce sperm in special containers called *antheridia* (singular, *antheridium*). The remaining, female plants form eggs in structures called *archegonia* (singular, *archegonium*). Sperms and eggs, being gametes, are haploid cells. Rain drops hitting the male plants, scatter the sperm to the females where sexual union of the gametes takes place within the archegonia. Zygotes, the products of that union, are retained within the archegonia.

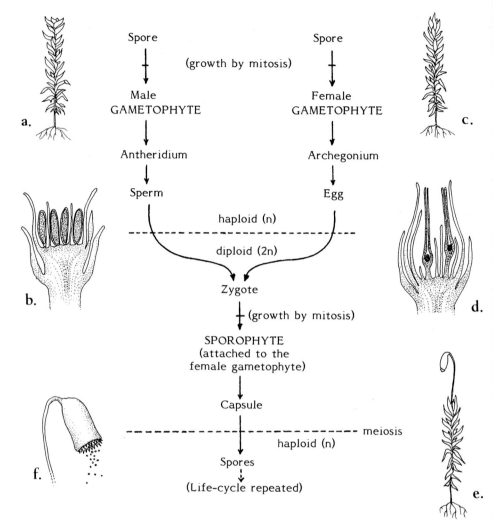

Spore   Spore

(growth by mitosis)

Male   Female
GAMETOPHYTE   GAMETOPHYTE

a.   c.

Antheridium   Archegonium

Sperm   Egg

haploid (n)

diploid (2n)

Zygote

b.   d.

(growth by mitosis)

SPOROPHYTE
(attached to the
female gametophyte)

Capsule

meiosis

haploid (n)

Spores

f.

(Life-cycle repeated)

e.

The life-history of a moss. Diagrams: a. Male gametophyte. b. Enlarged view of the top of the male gametophyte, showing four antheridia containing sperm. c. Female gametophyte. d. Enlarged view of the top of the female gametophyte, showing two archegonia with eggs. e. Asporophyte attached to the female gametophyte. f. A capsule discharges its spores.

Since the principal role of the green moss plants is to form gametes, they are called *gametophytes* (gamete-producing plants). The word ending, "-phyte" is derived from the Greek *phyton,* a plant.

With the formation of a zygote, a diploid cell, the moss life-cycle enters a completely different phase. The zygote undergoes repeated mitotic divisions to form a multicellular, diploid plant attached to the top of the female gametophyte. Since this new plant is not photosynthetic, it draws nourishment from the gametophyte during its growth into a slender stalk terminated by a tiny capsule. Within the capsule, spores are produced by the process of meiosis—haploid spores that, when dispersed by air currents, are ready to repeat the life-cycle.

The strange-looking diploid plants, balanced atop the green, female gametophytes, are called *sporophytes* since their special function is spore production. The sporophytes are familiar sights to all who have closely observed mosses in the field or garden.

The life-cycle of a moss entails the sequential development of two generations of plants—a haploid gametophyte, followed by a diploid sporophyte—even though the two plants live attached to one another. This amazing pattern of reproduction, called an *alternation of generations,* is the most common method of reproduction among plants, including ferns and Angiosperms.

The Life-cycle of a Fern

Fern leaves ("fronds") are among the most beautiful in the Plant Kingdom—the reason for these plants' special appeal to gardeners. The presence of vascular tissues in ferns, along with well-formed leaves, roots, and stems that are frequently rhizomatous, are indications of a significant evolutionary advance over the lowly mosses. So also is the predominant role played by the diploid sporophyte (compared with the gametophyte) in a fern's life history.

The fern plants gardeners know so well are sporophytes, as evidenced by the production of spores in tiny capsules clustered on the underside of the leaves. Such clusters, called *sori* (singular, *sorus;* from the Greek *soros,* a heap), are the familiar brown spots or stripes uniquely formed on fern leaves. Within the capsules, spores are produced by the process of meiosis.

As the capsules mature, dry, and split open, dust-like spores rain from the sori. Those spores fortunate enough to fall on moist soil, have the opportunity to grow into gametophytes. Fern gametophytes are heart-shaped plants, about ¼ inch (0.6 cm) broad in common species. They are photosynthetic and extremely fragile. During their brief existence, these haploid plants fulfill their destined sexual function by producing antheridia and archegonia, sperm and eggs, on the same tiny organism. Thus, to fertilize an egg, the male gametes have only a short distance to swim in a film of moisture on the gametophyte's lower surface.

During the initial growth of the zygote into a new sporophyte generation, the diploid plant is attached to the gametophyte—perhaps as a throwback to the union between the two generations in the mosses. But in ferns, the sporophyte's continued, vigorous

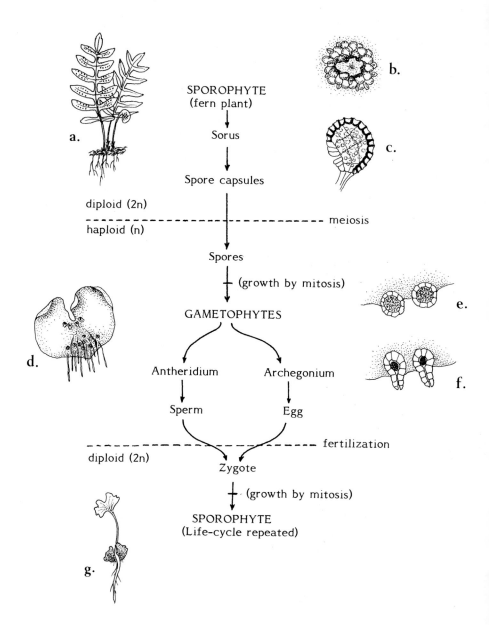

SPOROPHYTE
(fern plant)

a.

b.

c.

Sorus

Spore capsules

diploid (2n)
- meiosis
haploid (n)

Spores

(growth by mitosis)

GAMETOPHYTES

e.

d.

f.

Antheridium   Archegonium

Sperm   Egg

- fertilization
diploid (2n)

Zygote

(growth by mitosis)

SPOROPHYTE
(Life-cycle repeated)

g.

The life-history of a fern. Diagrams: a. The fern plant is a sporophyte. b. Close-up view of a sorus—a cluster of spore capsules. c. Enlarged view of a spore capsule. d. A gametophyte (sometimes called a "prothallus"). e. Enlarged view of two antheridia with sperm, attached to the under side of the gametophyte. f. Enlarged view of two archegonia with eggs. g. A young sporophyte, attached to the gametophyte.

growth soon overwhelms the gametophyte as the diploid generation establishes itself with roots, rhizomes, and successive sets of leaves.

The often unrecognized gametophyte generation of ferns, and the early development of the sporophytes, may be observed when fresh spores germinate in a flat of moist soil kept in a shady, cool place for several weeks. Close scrutiny is necessary to see the inconspicuous gametophytes whose pale-green color offers camouflage against the ground. The spores of some fern species germinate more readily than others. If successfully grown, there is no better opportunity to see "the other" generation in the reproductive cycle of a higher plant.

The Two Generations of Flowering Plants

As evolution progressed beyond the ferns, and culminated in the Angiosperms, further modifications were made to the system of alternation of generations.

The flowering plants produce two gametophytes: Of separate sexes; reduced to microscopic proportions; and dependent on the sporophyte for food and protection. Although these evolutionary remnants of the gamete-producing generation are barely recognizable as "plants", their place in the reproductive cycle is just as important as it is in primitive species. Contrary to popular belief, the plants we grow for their flowers, fruit, and seeds do not directly engage in sexual reproduction. They are sporophytes and, as such, produce spores by the process of meiosis—the singular purpose of the diploid generation.

In the anthers of a flower, for example, genetically selected, diploid cells undergo reduction division to form haploid *microspores* ("little spores", bearing the species' male characteristics). The content of each microspore divides by mitosis to form the two cells within the pollen (page 175). Further mitotic division of one of those cells to form two sperm, and extension of the other cell to form a pollen tube, is the full measure of the male gametophyte's growth. The latter events occur after pollination, and only when the pollen is supplied with food and growth hormones from the pistil's tissues.

When the germinated pollen is so viewed as a tiny gametophyte in the alternation of generations, it takes on added significance. For here is a legitimate, haploid, reproductive organism that, in the course of evolution, has been reduced to its bare essentials—two male gametes and a delivery tube. The capacity to photosynthesize and the development of antheridia have been lost. All that remains is a mere shadow of the pretty, green gametophytes of the primitive mosses.

Formation of the female gametophytes occurs within the

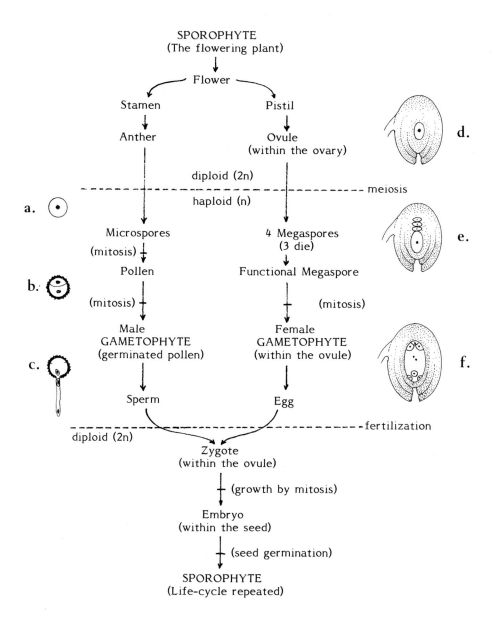

The life-history of a flowering plant. Diagrams: a. A microspore. b. Pollen grain. c. Germinated pollen—the microgametophyte generation. d. An ovule with the cell selected to undergo meiosis. e. Following meiosis, a functional megaspore with three aborted megaspores. f. An ovule containing the megagametophyte generation, including an egg cell (lower, center).

flowering plant's ovules. The process begins when a genetically selected cell divides by meiosis into four haploid *megaspores* ("large spores": bearing the species' female characteristics). Three of these megaspores die. The remaining functional megaspore undergoes repeated mitoses to form a collection of eight nuclei, divided between seven cells[1] that constitute the structure of the female gametophyte. One of the cells functions as an egg. Following fertilization, the resulting zygote grows into the seed's embryo (page 27) by way of many mitotic divisions. The embryo is an unborn sporophyte, waiting for the moment of seed germination when a new life-cycle is begun.

Similar to the male gametophytes, the female haploid plants are reduced to microscopic proportions; they are nourished by food from the sporophyte bearing them; and archegonia have been eliminated in the course of evolution. The bond between gametophyte and sporophyte is such that, in nature, existence of the female gamete-producing plant without the ovule's and ovary's protective walls is not feasible. And, except for apomictic embryo development (page 177), the formation of viable seeds by most flowering plants is only possible after the life-cycle has progressed through the intervening stage of the haploid generation.

In passing, it is worth noting that the reproductive cycle of Gymnosperms is comparable to that of the flowering plants, the principal difference being that the seeds are developed in cones. A pine tree, for example, is the sporophyte generation. And germinated pollen—the male gametophytes—grow inside the ovules where the tiny female gametophytes are harbored. The alternation of generation is as real in all higher plants as it is in primitive forms.

Chromosome Segregation during Meiosis

The occurrence of meiosis in a life-cycle is a crucial event, whether in the formation of spores, as in plants, or gametes, as in animals. During this process, members of paired chromosomes (*homologous pairs*) are randomly segregated into two haploid sets in separate cells. And with the chromosomes, genes controlling the expression of physical and physiological traits are also allocated to separate reproductive units.

Suppose a diploid sporophyte has only two pairs of chromo-

1. One of the female gametophyte's cells contains two haploid nuclei. This cell unites with the second sperm from the pollen tube to form a product having three sets of chromosomes—a triploid, or 3n, condition. Divisions of the triploid cell leads to the formation of a triploid *endosperm tissue* (page 176). In some species, variations in these events leads to endosperms having more than three chromosome sets.

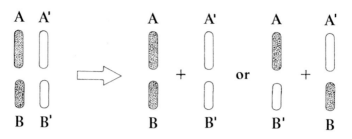

Segregation of chromosomes during meiosis. The homologous pairs at left are divided into haploid combinations at right.

A A B B	A A B B'	A A B'B'
A A' B B	A A'B B'	A A'B' B'
A' A' B B	A' A'B B'	A' A'B'B'

Possible chromosome recombinations, following the union of gametes into a zygote.

somes in each of its body cells. Let us label one pair A and A', the other B and B'. Following meiosis, haploid combinations of these chromosomes can only be: AB and A'B', or AB' and A'B (see the accompanying diagram). In the anther, meiosis distributes these chromosome combinations into the various microspores which, in turn, are transmitted to the pollen and sperm. However, only chance determines the chromosome combination present in the sperm that fertilizes an egg—a random selection, like the combination of numbers from the toss of two dice.

Likewise, in an ovule, meiosis creates the same pattern of chromosome segregations. But, again, chance determines which of the four megaspores survives to produce the female gametophyte and its egg (page 193).

After our hypothetical plant's gametes combine their chromosomes into a zygote's diploid nucleus, 9 chromosome patterns are possible (see the above table). When such a large number of combinations results from only four chromosomes, imagine the number of possible combinations obtainable with 40 chromosomes, a typical count for Angiosperm species.

Gene Segregations during Meiosis

Chromosome mixing during sexual reproduction takes on added significance when the thousands of genes borne by the chromosomes, and the traits they represent, are considered. But a further complexity is added, since genes have two modes of expression— they are either *dominant* or *recessive*. Dominant traits take precedence over the recessive when genes for both characteristics become mixed in an individual as a result of cross-breeding.

Consider the trait of plant height. Some hybrid varieties grow tall, others of the same species are genetic dwarfs. The gardener has a choice of growing "pole" or "bush" beans, for example. Tallness is frequently a dominant trait in plants, dwarfness is recessive.

For illustrative purposes, consider a plant having tall (T) genes on both members of an homologous chromosome pair (symbolized by a TT combination in the accompanying diagram). Pollen (and sperm donated by this plant to a sexual union) bear the tall trait since only the dominant genes are present. Consider another plant that is a genetic dwarf (t), having recessive genes on both of the chromosomes determining height (represented by a tt gene combination). All eggs produced by this plant carry the gene for dwarfness.

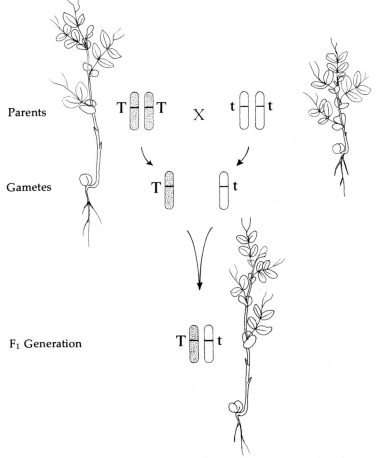

Parents

Gametes

F_1 Generation

A cross between a homozygous tall (TT) variety of pea and a dwarf (tt) results in heterozygous tall (Tt) progeny in the first filial (F_1) generation.

Following cross-pollination between these two plants, the fertilized egg contains the new gene combination *Tt,* as would every cell in the plant growing from the zygote. Would that plant be tall, dwarf, or an intermediate height?

Rarely do genes display partial expression. Most often, the presence of the dominant gene completely suppresses the expression of the recessive trait. Thus, the hybrid plant in question will grow tall and be called a *heterozygous* tall—a product of the pairing of the two different (hetero-) genes (*T* and *t*). The tall parent plant (*TT*) is a *homozygous* tall (homo-, same).

The visible expression of genes, in this case tallness, is called the *phenotype* of the plant (from the Greek, "to appear"); whereas the gene composition is called its *genotype.* Note that the occurrence of a recessive phenotype, a dwarf plant, is only possible when the double-recessive genotype (*tt*) is present.

The significance of these differences is that when two plants are cross-bred, both of which possess a recessive phenotype, all the offspring are guaranteed to possess that same trait. Similarly, all the progeny of two plants having homozygous dominant genotypes have the dominant characteristic. But when a heterozygous condition exists in either or both parents, the next generation can be expected to contain members with mixed phenotypes (see the accompanying diagram). Recessive genes, lurking in the parent's genetic makeup through several generations, eventually reveal their presence with the chance occurrence of the double-recessive gene combination.

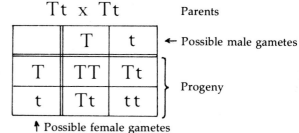

Tt x Tt   Parents

The results of a cross between two heterozygous tall plants. Progeny: 3 tall types to each dwarf (tt). Of the tall varieties, 2 are heterozygous tall (Tt) to each homozygous tall (TT).

← Possible male gametes

} Progeny

↑ Possible female gametes

The widespread presence of heterozygous gene combinations in plants is one of the principal reasons why hybrid progeny—the products of sexual reproduction—are not always true to their parent's form. The only way to guarantee genotypic uniformity among members of a population is to propagate them vegetatively by cuttings, grafts, etc.—techniques horticulturists regularly employ.

The inheritance of height is only one plant characteristic controlled by paired genes on homologous chromosomes. Thousands of other physical and physiological traits are so determined, including flower shape, fruit size and color, leaf form, response to photoperiod, and maturation rates. The genes for such features, located on different chromosomes, are thereby sorted and mixed into ever-changing combinations with each passing generation.

Each variety's complete, unique gene combination—its genetic blueprint—is so complex, that to duplicate it through cross-breeding is virtually impossible. Thus, when cultivars propagated by vegetative means are lost through human neglect, they can, in most cases, never again be re-created.

Consequences of Imperfection

In living organisms, the processes of mitosis and meiosis have occurred an infinite number of times, with astonishing regularity, since time immemorial. But, as with any biological system, cell divisions occasionally go wrong. The chromosomes may separate into incomplete sets during meiosis, resulting, for example, in gametes with an extra chromosome $(n+1)$, or with one missing $(n-1)$. Such a condition is called *aneuploidy* ("not good ploidy"). Most often, a plant is not at a disadvantage when it inherits an extra chromosome; in fact, it may result in the increased size of one or more of its organs, such as the broad leaves and globular fruits of the 'Globe' variety of Jimson Weed (*Datura stramonium*). However, the absence of a chromosome is fatal since too many genes necessary to the control of vital functions are missing.

A more common chromosomal aberration is called *polyploidy* ("poly-", many) in which cell nuclei possess three or more complete sets of chromosomes. Plants having *triploid* (3n) and *tetraploid* (4n) numbers of chromosomes abound in the typical garden. It has been estimated that ⅓ or more of Angiosperm species are polyploids, with an especially high representation among varieties of horticultural and agricultural interest. Of these, approximately 40% are dicots; 60% are monocots (particularly grasses). Many commercial plant catalogs feature the latest polyploid cultivars; some are grown from seed, others are propagated from vegetative organs.

Polyploidy frequently results in sterility (see below). Thus, species possessed of such a genetic makeup must reproduce by bulbs, corms, rhizomes, etc. (pages 108–111) to survive in nature. And it is through such methods, plus cuttings and grafts that horticulturists frequently propagate these unusual plants. Sterility appears in many ways, including: Disruptions in the development of pollen or pollen tubes; aborted embryos; or modifications in the structure of flower parts.

Polyploids are widely cultivated due to their large size and vigorous growth—endowed by the extra sets of genes in their cells.

Compared with their diploid counterparts, tetraploids, in particular, form larger leaves, flowers, and fruit. They possess increased food values, capacities for wood production, and overall stature. The tolerance of some polyploid species to environmental change has given them an advantage during natural selection, and greater flexibility for growth under a broad range of garden conditions.

The Origins of Polyploidy

Normally, sex cells contain the haploid number of chromosomes. But, sporadically, diploid gametes are formed when reduction fails to occur during meiosis (see the accompanying diagram). Thus, when a diploid gamete combines with a normal haploid sex cell, the resultant 3n zygote is the beginning of a triploid sporophyte generation.

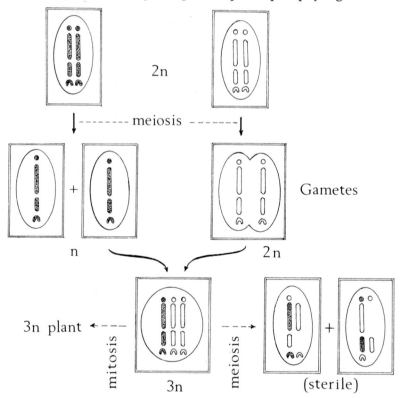

The origin of triploidy. Incomplete segregation of chromosomes during meiosis in one parent's cells (upper, right) ultimately leads to the formation of a diploid gamete. Union of that sex cell with a normal, haploid gamete (upper, left) results in a triploid zygote. The zygote can undergo mitotic divisions to form a 3n plant; but when the plant attempts meiosis, irregular segregation of the chromosomes leads to sterility.

Vegetative propagation is possible in a 3n plant since mitotic cell divisions simply replicate the triploid chromosomes in each newly formed cell.

The third set of chromosomes may confer hybrid vigor to the now growing plants, but reduced fertility is virtually assured when what seems to be normal reproduction occurs between triploids. During meiosis, the irregular distribution of the three sets of chromosomes to the spores, followed by their passage to the gametophytes and gametes, eventually results in incomplete chromosome pairing when the sex cells unite. For subsequent meiosis to occur, paired homologous chromosomes must be present.

Fusion of two 2n gametes leads to the formation of a tetraploid

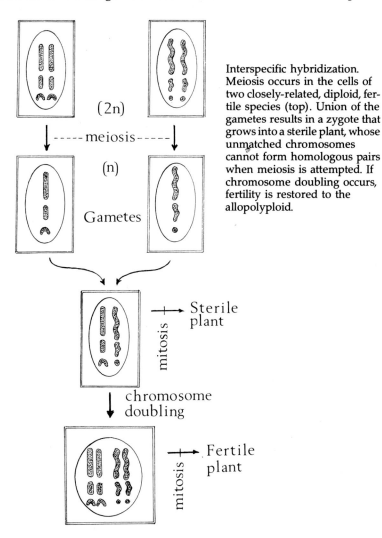

(2n)

----- meiosis -----

(n)

Gametes

Interspecific hybridization. Meiosis occurs in the cells of two closely-related, diploid, fertile species (top). Union of the gametes results in a zygote that grows into a sterile plant, whose unmatched chromosomes cannot form homologous pairs when meiosis is attempted. If chromosome doubling occurs, fertility is restored to the allopolyploid.

mitosis ──→ Sterile plant

chromosome doubling

mitosis ──→ Fertile plant

(4n) plant. The gametes may be donated by two plants of the same species (*intraspecific hybridization*), or by plants of different, but closely-related species (*interspecific hybridization*). Although it was mentioned in a previous section (page 77) that different species are generally recognized by their inability to cross-breed, polyploids derived from two or more species (*allopolyploids*—"allo-", different) do occur. Thus, allopolyploidy is one of the principal ways in which genetic barriers between species are broken down, resulting in new gene combinations that may confer selective advantages to the offspring. Such crosses are also of interest to mankind, since new phenotypic variations, typically arising in allopolyploids, have significant potentials for economic exploitation. Varieties of apple, grape, loganberry, maize, rice, strawberry, rose, dahlia, chrysanthemum, gladiolus, and several species of orchid are among the different cultivars developed in this manner.

Sterility in interspecific hybrids is due to the inability of unlike chromosomes, from parents of diverse species, to find partners and form homologous pairs, necessary for later meiosis. Sometimes, however, all such chromosomes double, thus providing themselves with identical mates for normal pairing (see diagram). The resulting allopolyploid plant is completely fertile since it simply functions as a diploid with a large number of homologous chromosomes.

The ancestors of the garden dahlia (*Dahlia pinnata*) are believed to have had a diploid number of 16 chromosomes (2n = 16). Hybridization probably brought about the development of two groups of tetraploid species (2n = 32). The garden species is an interspecific octaploid hybrid (2n = 64) of those groups formed by the chromosome-doubling described above. The garden dahlia is, incidentally, a highly fertile species. Other hybrids that have been traced to crosses between two or more species include the domestic plum (*Prunus domestica*), leaf mustard (*Brassica juncea*), and commercial tobacco (*Nicotiana tabacum*).

Custom-made Plants of the Future

The horticulturist's and agronomist's traditional methods of plant breeding are fraught with problems related to the games of chance Nature plays during sexual reproduction. For example, the plant breeder has little control over chromosome segregations in meiosis, or over gene sorting at the time of gamete union. And to obtain the unique benefits of polyploidy, he is dependent on the outcome of the random, sporadic misbehavior of cells at various stages in the reproductive cycle. Colchicine, a substance extracted from the autumn crocus (*Colchicum autumnale*), is one of many chemicals used to artificially induce polyploidy, but the results are as unpredictable as when polyploidy occurs by natural means.

The purpose of selective breeding is to gather certain genes into hybrid cultivars that mankind deems useful or aesthetically pleasing. Until recent times, genetic traits that breeders transmitted between plant generations were restricted to those already present in the gene pool of the species, or closely-related taxa. And genes were only carried on whole, compatible chromosomes in the company of other units of inheritance, desirable or otherwise.

With the discovery of how genetic information is contained within the elaborate molecular structure of DNA (page 18), and the recent development of ingenious techniques of gene manipulation, plant breeding has entered a new era. It is now possible to remove individual genes from cell nuclei, and splice them into the chromosomes of other plants. Gene transfers can be carried out between species from different families, and even between lower forms and flowering plants. Such is the marvel of genetic engineering.

Most genes inserted into an organism's chromosomes function in a normal manner, simply directing the cell to perform a new set of biochemical activities. And when the newly-designed chromosomes engage in mitosis or meiosis, they pass the introduced genes to subsequent generations of cells. Furthermore, when genetically-engineered plants are propagated by cloning (page 72), the painstaking procedures of cross-breeding by way of sexual reproduction are completely by-passed.

As botany enters the 21st century, mass production of custom-made plants is envisioned: Plants possessing the remarkable resistance of some native species to pathogens and harmful insects; agricultural crops designed to withstand drought and saline soils, with genes transposed from desert species that learned such tricks long ago, in the hard way; plants with improved nutritional value, productivity, stronger fibers, or straighter wood are certain to be developed. Ordinary and easily grown species will be turned into drug factories with genes currently possessed by only the exotic and rare. The abnormally high photosynthetic yields of a few species will become common to all that are of economic importance to mankind. And since gene sharing knows no taxonomic boundaries, the important process of nitrogen-fixation (page 114) need no longer be the exclusive work of a few bacteria and blue-green algae.

Perhaps, somewhere along this assembly line, a blue rose or a red delphinium will emerge. Whether that is a good idea or not, is for the reader to decide.

Epilogue

In these pages, we have ventured into a realm familiar to gardeners, although not always well understood. We have journeyed far into the Plant Kingdom. Words and pictures have formed interwoven paths of ideas as we traversed millions of years of evolution and, occasionally, stepped from our gardens into earth's far-flung corners.

We have explored the microscopic world of plant cells and tissues; climbed inside roots, stems, leaves, and flowers; and traced the connections unifying such diverse parts into whole, living organisms. In the course of our travels, we have paused to wonder at how plants are made, to follow the progress of their growth, and attempt to comprehend some of their mysteries.

We have delved into the intricacies of reproduction, inheritance, photosynthesis, and natural selection. And have witnessed the power of plants to survive by use of clever defense systems and their uncanny ability to adapt to the contingencies continually challenging each species' existence.

Our expedition has been made possible by virtue of the pioneering work of countless scientists, gardeners, and other plant lovers who, over hundreds of years, have unravelled and created concepts of what plants are and how they function—a picture that continues to grow, and never ceases to inspire awe and enthusiasm for these delightful creatures.

If this introduction to botany has opened new vistas for the reader's thoughts, encouraged a closer look at the plants in his or her surroundings, and brought a greater sense of appreciation for Nature's splendid handiwork, it has served a good purpose. Now let it be a guide as you embark on many more journeys of discovery.

Further Reading

Ayensu, Edward S., and V. H. Heywood, G. L. Lucas, R. A. Defilipps, *Our Green and Living World.* Cambridge University Press, New York, 1984.

Beazley, Michael (Editor), *The International Book of the Forest.* Simon & Schuster, New York, 1981.

Heywood, V. H. (Editor), *Flowering Plants of the World.* Oxford University Press, Oxford, 1979.

Heywood, V. H., and S. R. Chant (Editors), *Popular Encyclopedia of Plants.* Cambridge University Press, New York, 1982.

Hohn, Reinhardt, *Curiosities of the Plant Kingdom.* Universe Books, New York, 1980.

Lehane, Brendan, *The Power of Plants.* McGraw-Hill, New York, 1977.

Meeuse, Bastiaan, and Sean Morris, *The Sex Life of Flowers.* Facts on File, New York, 1984.

Milne, Lorus, and Margery Milne, *Living Plants of the World.* Random House, New York, 1969.

Moore, David M. (Editor), *Green Planet: The Story of Plant Life on Earth.* Cambridge University Press, New York, 1982.

Paturi, Felix R. *Nature, Mother of Invention: The Engineering of Plant Life.* Harper & Row, New York, 1976.

Proctor, John, and Susan Proctor, *Color in Plants and Flowers.* Everest House, New York, 1978.

Rost, T. L., and M. G. Barbour, R. M. Thornton, T. E. Weier, C. R. Stocking, *Botany, a Brief Introduction to Plant Biology.* Wiley, New York, 1984. (An excellent, college-level text.)

Went, Frits W., *The Plants.* Time Inc., New York, 1963.

Glossary-Index

Abscisic acid. A growth-inhibiting hormone, pp. 127, 130.

Abscission. The controlled separation of leaves, flowers and fruit from plants, pp. 92, 127, 148.

Abscission zone. A layer of cells at the base of a leaf petiole, flower, or fruit stalk, the weakening of which causes the organ to separate from the plant, p. 128.

Actinomorphic flower. A flower possessing radial symmetry: Any cut through the center divides the flower into two equal parts, p. 165.

Adventitious root. A root arising in an unexpected position, such as from a leaf, pp. 101, 105, 110, 129.

Adventitious shoot. A shoot arising in an unusual position, such as from the side of a root, pp. 101, 111.

Aerial root. A root emerging above soil level, p. 106.

After-ripening. A maturation process in seeds of particular species after dispersal, required for germination, p. 32.

Aggregate fruit. A group of small fruits derived from several ovaries within a single flower, p. 179.

Alkaloid. A nitrogen-containing compound, frequently used as a chemical defense by plants, p. 94.

Allelopathy. Release of chemicals by a plant to discourage the growth of other plants near it, pp. 33, 100.

Allopolyploid. A hybrid arising from the combination of chromosomes from two different species, p. 200.

Alternation of generations. The sequence of a haploid gametophyte and a diploid sporophyte during the course of a life-cycle, pp. 189, 191.

Aneuploidy. A condition in which chromosome numbers are not in exact multiples of the haploid set; having extra or missing chromosomes within a nucleus, p. 197.

Angiosperm. A member of a class of plants characterized by the formation of flowers, and seeds in fruits, pp. 13, 75, 160, 191.

Annual. A plant completing its life-cycle within a single growing season, pp. 16, 83, 84, 134.

Annual ring. A cylinder of secondary xylem added to the wood in a single growing season, p. 60.

Anther. The pollen-bearing part of a stamen, pp. 164, 191.

Antheridium. The male sex organ of plants other than Gymnosperms and Angiosperms, pp. 187, 189.

Anthocyanin. A water-soluble pigment, varying from red to blue in color, p. 151.

Apical bud. A bud at the tip of a stem, pp. 39, 44.

Apical dominance. The inhibition of axillary bud growth by the apical bud, p. 129.

Apical meristem. A region of actively dividing cells at the tip of a growing root or stem, pp. 23, 37, 38.

Apomixis. Development of a viable seed without fusion of gametes, p. 177.

Archegonium. The female sex organ of plants, other than Angiosperms, pp. 187, 189.

Autotrophic nutrition. A form of nutrition in which complex food molecules are produced by photosynthesis from carbon dioxide, water, and minerals, pp. 32, 137, 154.

Auxin. A plant hormone that principally controls cell elongation, pp. 121, 123, 129.

Axil. The angle between the upper surface of a leaf and the stem to which it is attached, p. 39.

Axillary bud. A bud located in an axil at the base of a leaf, pp. 39, 108, 110, 129.

Axillary bud primordium. An immature axillary bud, p. 39.

Bark. All the tissues, collectively, formed outside the vascular cambium of a woody stem or root, pp. 43, 57, 59.

Biennial. A plant completing its life-cycle within two growing periods, pp. 16, 83, 132.

Blade. The flattened part of a leaf, pp. 45, 47, 67.

Bolting. The rapid growth of a stem prior to flowering, p. 132.

Bract. A modified leaf arising below a flower or inflorescence, (photograph) p. 166.

Branching, of roots, pp. 38, 66; of stems, pp. 39, 42, 59.

Bud scale. A modified leaf protecting a bud, pp. 44, 84, 131.

Bud scale scar. See terminal bud scale scar.

Bulb. A short, flattened stem bearing fleshy, food-storage leaves, pp. 108, 131, 197.

Buttress root. An enlarged, above-ground root giving support to a tree trunk, p. 105.

Callose. A plant substance created and deposited in the pores of phloem sieve plates, especially in response to injury, p. 92.

Callus. A corky tissue developed by woody species to cover wounds, p. 91.

Calyx. Collectively, all of the sepals in a flower, p. 162.

Cambium. See vascular cambium, cork cambium.

Capillary water. Water held in the tiny spaces between soil particles or between plant cells, p. 149.

Carnivorous plant. See insectivorous plant.

Carotene. An orange-yellow pigment located in the chloroplasts, p. 151.

Cell. The smallest, independently alive unit from which plants and animals are constructed, pp. 18, 71.

Cellular respiration. The chemical breakdown of food substances, resulting in the liberation of energy, pp. 20, 31, 154, 156.

Cellulose. A plant substance forming a part of the structure of cell walls, pp. 21, 74, 155.

Cell wall. The outer covering of a plant cell, pp. 20, 137.

Chelate. An organic substance to which metals such as iron are bound and from which they are released, p. 151.

Chlorophyll. A green plant pigment located in chloroplasts, pp. 20, 146, 151.

Chloroplast. A cellular body in which photosynthesis occurs, pp. 19, 151.

Chlorosis. An abnormal yellowing of leaves due to a reduced chlorophyll content, p. 146.

Chromosome. A thread-like structure bearing genes in a cell nucleus; each chromosome consists of two chromatids formed by the chromosome's longitudinal division, pp. 23, 146, 185, 186, 193, 197.

Cladode. A flattened stem performing the function of a leaf (e.g. a cactus pad), p. 107.

Clay. An inorganic soil component having particles less than 0.002 mm diameter, pp. 149, 150.

Cleistogamy. The development of viable seed from unopened, self-pollinated flowers, p. 174.

Clones. Genetically identical organisms produced vegetatively from a single parent, pp. 72, 111, 201.

Cold hardening. The process whereby some species prepare for seasonal periods of low temperatures, p. 144.

Companion cell. A phloem cell containing a nucleus, adjacent to a sieve tube, p. 72.

Complete flower. A flower having all of the normal flower parts, p. 162.

Composite head. An inflorescence composed of many tightly packed, small, ray and disc flowers, pp. 168, 170.

Compound leaf. A leaf in which the blade is divided into separate leaflets, p. 47.

Contractile root. A thickened root serving to pull a corm, bulb, or rhizome deeper into the soil, p. 108.

Cork. The protective, outer tissue of the bark, pp. 44, 57, 67, 84.

Cork cambium. A layer of cells in the bark giving rise to the cork; a lateral meristem, pp. 57, 67, 84.

Corm. A short, swollen, underground stem in which food is stored, pp. 109, 197.

Cormel. A small, undeveloped corm, p. 110.

Corolla. Collectively, all the petals in a flower, p. 163.

Corona. A trumpet-like outgrowth of petals, p. 165.

Cortex. The tissue in roots and stems immediately inside the epidermis, pp. 54, 65.

Cotyledon. A seed leaf; a food storage structure in seeds, pp. 27, 29.

Critical photoperiod. The maximum day length a short-day plant and the minimum day length a long-day plant require to initiate flowering, p. 134.

Cross-pollination. The transfer of pollen to a flower on another plant, p. 173.

Cultivar. A cultivated variety, produced by horticultural techniques, p. 184.

Cuticle. A waxy layer on the outside of leaves, herbaceous stems and fruits, pp. 54, 69, 86.

Cutin. The waxy substance forming a cuticle layer, pp. 54, 69, 90.

Cytokinin. A plant hormone primarily stimulating cell division, p. 127.

Cytoplasm. The living protoplasm of a cell, excluding the nucleus, pp. 18, 137.

Cytoplasmic membrane. The membrane enclosing the cytoplasm, pp. 18, 137.

Day-neutral plant. A plant in which flower formation is not controlled by photoperiod, p. 134.

Deciduous. Shedding all of the leaves in one season, pp. 34, 127.

Defoliant. A synthetic chemical causing leaves to be prematurely shed, p. 130.

Determinate growth. Growth to a genetically pre-determined size, p. 15.

Diageotropic. Horizontal growth of a plant part, p. 123.

Dicot. A member of a subclass of Angiosperms characterized by having two cotyledons in their seeds, p. 29.

Differentiation. The process whereby parenchyma cells undergo morphological and physiological change in order to become specialized in function, p. 52.

Diffuse root system. See fibrous root system.

Diffuse secondary growth. Thickening of a plant organ (e.g. a palm trunk) by scattered cell divisions, rather than growth originating from cambium tissues, p. 62.

Dioecious. Having male and female sex organs on separate individuals, p. 173.

Diploid. Having two sets of chromosomes, p. 186.

Disbudder. A synthetic chemical causing the shedding of immature flower buds, p. 130.

Disc flower. A small, tubular flower at the center of a composite head, p. 169.

DNA. Deoxyribonucleic acid. The substance of which genes are made; the carrier of genetic information in cells, pp. 18, 23, 146, 201.

Dormant, dormancy. A state of reduced cellular activity, pp. 26, 84, 131.

Dominant species. The most abundant species in a plant community, and the one most impacted by the environment. p. 100.

Dominant trait. A characteristic determined by a gene masking the expression of a comparable, but recessive gene, p. 194.

Drip tip. A pointed leaf tip helping to drain water from the leaf surface, p. 106.

Drip zone. The area of soil around a tree occupied by root tips and into which water drips from the leaf canopy, p. 37.

Egg. A female sex cell, pp. 160, 176, 187, 189, 193.

Embryo. An immature plant within a seed, pp. 27, 177.

Endodermis. A layer of cells in roots between the cortex and vascular tissues, pp. 66, 140.

Endosperm. Food-storage tissue in seeds, pp. 29, 176, 193.

Enzyme. A protein molecule functioning as a chemical catalyst in a biochemical reaction, pp. 93, 146.

Epidermal hair. A filament of cells arising from an epidermal cell, pp. 54, 86, 142.

Epidermis. The outer layer of cells on an herbaceous plant organ, pp. 54, 65, 140.

Epigeous germination. Seed germination in which the cotyledons are raised above the soil surface, p. 29.

Epiphyte. A plant growing upon another plant for physical support, p. 104.

Ethylene. A gaseous plant hormone produced in abundance by ripening fruits and damaged tissues, pp. 127, 128, 130.

Etiolation. The condition of a plant when grown in darkness; its stem is pale and elongated, the leaves are undeveloped, p. 120.

Evergreen. A woody perennial plant bearing leaves throughout the year, pp. 44, 86, 128.

Fermentation. The partial breakdown of food molecules to yield ethyl alcohol, CO_2 and energy; occurs in the absence of oxygen, p. 156.

Fiber. A long, thick-walled cell dead at maturity, p. 62, 73.

Fibrous root system. A highly-branched, spreading root system, pp. 35, 107.

Field capacity. See water-holding capacity.

Filament. The stalk of a stamen, bearing an anther, p. 164.

Flower. The reproductive branch of an Angiosperm plant, pp. 133, 159, 162.

Food. An organic substance providing energy and body-building materials; especially carbohydrates, fats and proteins, p. 153.

Food conduction, pp. 55, 58, 69, 72.

Food storage, pp. 36, 66, 108.

Freezing, prevention of in cells, pp. 86, 144.

Fruit. A mature ovary, pp. 13, 162, 164, 176, 177.

Fruit ripening, p. 127.

Gamete. A sex cell; sperm or egg, pp. 160, 188, 189, 198.

Gametophyte. A haploid, gamete-producing plant in the alternation of generations, pp. 188, 189, 191.

Gas exchange, pp. 44, 69, 70, 156.

Gene. A unit of genetic inheritance, pp. 15, 20, 23, 185, 194, 201.

Genotype. The genetic constitution of an organism, p. 196.

Genus. A taxonomic category containing related species, p. 183.

Geotropism. Growth of a plant organ in response to gravity, p. 122.

Germination. The beginning of growth of a seed, spore or pollen grain, pp. 30, 32–34.

Germination inhibitor. A chemical substance preventing seed germination, p. 33.

Gibberellin. A plant hormone regulating several processes including internode elongation and cell enlargement, pp. 123, 130, 131, 177.

Glaucous. Smooth and having a waxy bloom, p. 54.

Graft. The union of a piece of one plant to another, established plant, pp. 60, 91.

Granum. A stack of plate-like, pigment-containing structures in a chloroplast, p. 151.

Gravitropism. See geotropism.

Growth retardant. A chemical substance slowing or inhibiting plant growth, p. 130.

Guard cell. One of a pair of cells surrounding a stoma, p. 71.

Gum. A sticky, water-soluble plant secretion that hardens on exposure to air, p. 92.

Guttation. Exudation of droplets of water, most often from leaf margins, as the result of water movement up a plant due to root pressure, p. 140.

Gymnosperm. A member of a class of plants forming seeds in an exposed condition, frequently in cones, pp. 13, 73, 160, 172, 182, 193.

Haploid. Having one set of unpaired chromosomes, p. 186.

Haustorium. An organ produced by a parasite that penetrates and absorbs water and nutrients from the host's tissues, p. 111.

Heartwood. The central, dark-colored portion of secondary xylem in a tree trunk, p. 60.

Hemiparasite. A parasite that invades its host to obtain only water and mineral nutrients, p. 112.

Herbaceous. Soft, green and containing little woody tissue, pp. 30, 54.

Herbicide. Any chemical that, when applied to a plant, inhibits growth or kills, p. 130.

Heterotrophic nutrition. A form of nutrition in which the organism depends on organic substances as a food source, as is the case with humans, pp. 31, 137.

Heterozygous. Having both dominant and recessive genes for a particular characteristic on homologous chromosomes, p. 196.

Homologous chromosomes. Matching chromosome pairs, pp. 193, 200.

Homozygous. Having identical genes on homologous chromosomes, p. 196.

Hormone. An organic substance produced in small amounts and transported to sites where it controls growth and developmental processes, p. 120.

Host. A plant or animal harboring a parasite, p. 111.

Humus. Organic matter in the soil derived from the decomposition of plant and animal remains, p. 149.

Hybrid. The offspring of two plants of the same or closely related species differing in one or more genes, pp. 72, 77, 196, 200.

Hybrid vigor. The increased vigor, size, and fertility of a hybrid compared with its parents, p. 77, 199.

Hypocotyl. The part of a seedling between the roots and the place of attachment of the cotyledons, p. 29.

Hypogeous germination. Seed germination in which the cotyledons remain below the soil surface, p. 29.

Imbibition. The process of water absorption by a dry substance or structure, causing it to swell, p. 31.

Incomplete flower. A flower lacking one or more of the normal flower parts, p. 164.

Indeterminate growth. Growth to an indefinite size, p. 16.

Inflorescence. A shoot bearing clusters of flowers, p. 168.

Insectivorous plant. A plant that captures and digests insects as a source of nitrogen, pp. 115, 126.

Intercalary meristem. A meristem located between non-dividing tissues such as at the base of a leaf, p. 49.

Internode. The segment of a stem between two nodes, pp. 39, 123.

Lateral bud. See axillary bud.

Lateral meristem. A region where cells divide, located along the length of a stem or root (e.g. vascular and cork cambia), pp. 23, 57.

Latex. A thick, white, fluid secretion of many plant species, p. 92.

Laticifer. A cell producing latex, p. 92.

Layering. A method of plant propagation in which adventitious roots are developed on an intact plant, before the rooted section is removed, p. 129.

Leaf. An outgrowth of a stem; the principal organ of photosynthesis, pp. 45, 67, 115.

Leaf arrangements and shapes. (Diagrams) pp. 41, 48.

Leaflet. A portion of the blade of a compound leaf, p. 47.

Leaf primordium. An immature leaf, located at a stem tip, p. 39.

Leaf rosette. A group of leaves radiating from a short stem, pp. 106, 132.

Leaf scar. A scar left on a stem after a leaf has fallen, p. 44.

Leaf tendril. A modified leaf or leaf part used as a grasping organ, p. 103.

Lenticel. A small, gas-exchange opening in the cork of a woody stem p. 44.

Liana. A long-stemmed, woody, climbing plant growing from the ground into the tree canopy of tropical forests, p. 104.

Lignin. A tough, durable plant substance deposited in cell walls, especially in wood, p. 21.

Light absorption, pp. 45, 67, 100, 151, 154.

Living stone. A species of succulent plants camouflaged to look like a small rock, p. 89.

Loam. A mixture of sand, silt, and clay, p. 149.

Long-day plant. A plant flowering in response to day lengths exceeding its critical photoperiod, p. 134.

Macronutrient. A mineral required by plants and animals in relatively large quantities, p. 145.

Megaspore. A spore that develops into a female gametophyte, p. 193.

Meiosis. A cell divisional process in which the chromosome number is reduced by half, pp. 186, 198.

Membrane. A thin, sheet-like structure composed of protein and fats surrounding the cytoplasm, organelles, and other cell structures, pp. 18, 137, 146.

Meristem. A region where cells actively divide, p. 23.

Mesophyll. The parenchyma tissue of a leaf between the upper and lower epidermis, including palisade and spongy cells, p. 67.

Metabolism. The sum of the biochemical processes of a living organism, pp. 15, 17, 93, 158.

Microfibril. A fine tread of cellulose in a cell wall, pp. 21, 22.

Micronutrient. A mineral required by plants and animals in relatively small quantities, p. 145.

Microspore. A spore that develops into a male gametophyte, p. 191.

Middle lamella. A layer of pectin binding two adjacent cell walls, pp. 20, 128.

Mitochondria. Cellular bodies in which cellular respiration occurs, pp. 20, 154.

Mitosis. A cell divisional process in which the chromosomes are duplicated, pp. 23, 186.

Monocot. A member of a subclass of Angiosperms characterized by the presence of one cotyledon in the seeds, pp. 29, 48, 62.

Monoecious. Bearing separate male and female flowers on the same plant, p. 173.

Molecule. A chemically bonded group of atoms, p. 52.

Multiple fruit. A cluster of mature ovaries from several flowers on a single stem, p. 179.

Mutation. An induced, inheritable change in the structure of a gene, p. 77.

Mycorrhiza. An association between a fungus and the roots of a higher plant, p. 113.

Nastic movement. A movement of a plant part (e.g. a leaf) not caused by an external stimulus, p. 126.

Natural selection. The action of the environment on organisms such that those better able to survive environmental stress are more likely to reproduce and perpetuate their species, pp. 76, 87, 198.

Necrosis. The death of a plant tissue, p. 148.

Nectar. A sugary fluid secreted in some flowers, p. 166.

Nectar guide. A contrasting color pattern in a flower that guides a pollinator to the nectar, p. 168.

Nectary. A gland secreting nectar, p. 166.

Nitrogen cycle. The circulation of nitrogen between the environment and living organisms, p. 114.

Nitrogen-fixation. The conversion of atmospheric nitrogen (N_2) into organic nitrogen compounds by a limited number of microorganisms, p. 114.

Node. The segment of a stem to which leaves and axillary buds are attached, p. 39.

N-P-K ratio. The relative proportions of nitrogen, phosphorus and potassium in a fertilizer, p. 148.

Nucleus. The body within a cell controlling its activities, including inheritance, p. 18, 20.

Nutrient uptake, pp. 35, 113, 144.

Organ. A part of a plant, composed of different tissues, that acts as a functional unit, p. 53.

Organelle. A cell structure performing a specific function, p. 19.

Organic. Referring to substances containing, at least, both carbon and hydrogen, pp. 111, 149.

Organism. A living plant or animal, p. 53.

Osmosis. The diffusion of water across permeable cell membranes which select for or against specific substances, pp. 31, 137.

Ovary. The basal portion of a pistil that becomes a fruit, p. 164.

Ovule. An immature seed, pp. 164, 175.

Palisade cell. A photosynthetic cell directly beneath the upper leaf epidermis, p. 67.

Palmate venation. A vein pattern in which the major veins radiate from one point, p. 48.

Palmately compound leaf. A leaf in which the leaflets radiate from one point, p. 47.

Panicle. A highly branched inflorescence, (diagram) p. 170.

Parallel venation. A vein pattern in which the veins are arranged parallel to each other, p. 48.

Parasite. A plant or animal obtaining food from another living organism, frequently to the latter's detriment, p. 111.

Parenchyma. A thin-walled, undifferentiated cell, p. 72.

Parthenocarpy. Development of a fruit without pollination, fertilization or seed development, p. 177.

Pectin. A substance in cell walls binding cells together, pp. 20, 128, 148.

Pedicel. The stalk of an individual flower in an inflorescence, p. 162.

Perennial. A plant living through several growing seasons, pp. 16, 83, 84, 134.

Perianth. All the sepals and petals in a flower, p. 163.

Pericarp. The fruit wall; derived from the ovary wall, p. 177.

Pericycle. A root tissue giving rise to branch roots, p. 66.

Petal. A frequently flattened, conspicuously colored flower part, p. 163.

Petiolate leaf. A leaf in which the blade is attached to the stem by a petiole, p. 45.

Petiole. A leaf stalk, p. 45.

pH. A measure of relative acidity or alkalinity, p. 150.

Phenotype. The physical appearance of an organism, p. 196.

Phloem. The food-conducting tissue of plants, pp. 52, 55, 58, 66, 67, 72, 92.

Photoinduce. To initiate a physiological process as a result of being subjected to a particular photoperiod, p. 134.

Photoperiodism. The initiation of flowering in response to relative lengths of day and night, pp. 80, 133, 162.

Photosynthesis. The process in which light energy is used to form foods from carbon dioxide (CO_2) and water (H_2O), pp. 13, 93, 137, 154.

Phototropism. Curvature of a plant organ in response to light, p. 120.

Phytoalexin. A chemical produced by a plant to inhibit the growth of pathogens, p. 97.

Phytotoxin. A plant product having toxic effects on herbivores and other invasive organisms, p. 95.

Pilose. Having long, soft hairs, p. 89.

Pinnate venation. A vein pattern in which the major veins are arranged in rows on each side of the midrib, p. 48.

Pinnately compound leaf. A leaf in which the leaflets are arranged on both sides of a common axis, p. 47.

Pistil. The female part of a flower, p. 164.

Pit. A small opening in a cell wall, p. 73.

Pith. A parenchyma tissue at the center of a stem, p. 54.

Plagiotropic. Growth of a branch at an angle, other than 90°, to the vertical plant axis, p. 123.

Plant growth regulator. See hormone.

Plasmodesmata. Fine strands of cytoplasm that pass through cell walls, connecting adjacent cells, p. 20.

Plasmolysis. Shrinkage of cytoplasm away from the cell wall as a result of excess water loss, p. 139.

Pollen. A structure that develops from a microspore in Angiosperms and Gymnosperms to become a male gametophyte, pp. 164, 175, 191.

Pollen tube. An outgrowth from a pollen grain conveying the sperm to the female gametophyte, pp. 175, 191.

Pollination. Pollen transfer from an anther to a stigma or, in Gymnosperms, from a male cone to a female cone, pp. 165, 171.

Polyploid. Having three or more sets of chromosomes per cell, pp. 197, 198.

Prickle. A hard, pointed epidermal outgrowth on some species' stems and leaves, p. 88.

Primary growth. Growth arising from cellular activities in apical meristems, pp. 23, 37, 39.

Primary phloem. Food-conducting tissue formed by growth activities originating in apical meristems, pp. 56, 66.

Primary tissue. A tissue formed during primary growth, p. 56.

Primary wall. The first layer of cellulose laid down during development of a new cell wall, p. 21.

Primary xylem. Water-conducting tissue formed by growth activities originating in apical meristems, pp. 56, 66.

Prop root. A supportive root growing from an above-ground stem, p. 105.

Protoplasm. The living substance of cells, including cytoplasm and nucleus, p. 18.

Pubescent. Having short hairs, pp. 54, 89.

Raceme. An inflorescence in which flowers are borne on short stalks on an elongated stem, (diagram) p. 169.

Radicle. An embryonic root, p. 27.

Raphide. A needle-shaped crystal of calcium oxalate in certain species' cells, p. 97.

Ray flower. One of several small flowers often forming a ring around the disc flowers in a composite head, p. 169.

Receptacle. The enlarged end of a flower stalk to which the flower parts are attached, pp. 162, 179.

Recessive trait. A genetic characteristic the expression of which is masked by a comparable but dominant gene, pp. 194, 196.

Resin. A viscous, protective secretion of many conifers that is insoluble in water and hardens on contact with air, p. 92.

Resin canal. A resin-containing tube, p. 92.

Respiration. See cellular respiration.

Reticulate venation. A net-like vein pattern in some leaves, p. 48.

Rhizome. An underground, horizontal stem, pp. 101, 110, 197.

Ribosome. A cellular particle; the site of protein synthesis, p. 20.

Ripeness-to-flower. The minimal vegetative size a plant must achieve before it is capable of flowering, pp. 134, 148.

Root. Generally the underground portion of a plant; an organ anchoring the plant to the soil and absorbing water and minerals, pp. 16, 35, 37, 105, 148.

Root cap. A protective cover over a root tip, pp. 37, 125.

Root hair. A hair-like projection of a root's epidermal cell, pp. 38, 65.

Root nodule. A small swelling on a root resulting from invasion by nitrogen-fixing bacteria, p. 114.

Root pressure. The pressure developed by living cells in a root forcing water up the xylem, p. 140.

Root tuber. An enlarged, food-storage root bearing adventitious shoots, p. 111.

Runner. A horizontal stem growing above ground that may form roots at its tip or at nodes, p. 101.

Sand. An inorganic soil component the particles of which range between 0.02 and 2 mm in diameter, p. 149.

Saprophyte. An organism obtaining food from dead organic matter, p. 111.

Sapwood. The outer, light-colored, water-conducting region of secondary xylem, p. 60.

Scarify. To scratch or etch a thick seed coat to improve water uptake, p. 26.

Scion. A plant part inserted into a root stock during grafting, p. 60.

Sclereid. See stone cell.

Secondary growth. Growth resulting from the activities of lateral meristems (vascular and cork cambia), pp. 23, 58, 67.

Secondary phloem. Food-conducting tissue formed by the vascular cambium, pp. 56, 67.

Secondary product. A biochemical product other than substances used in major metabolic pathways such as photosynthesis and respiration, p. 93.

Secondary wall. The portion of a cell wall laid down inside the primary wall, p. 21.

Secondary xylem. Water-conducting tissue formed by the vascular cambium, pp. 56, 67.

Seed. A reproductive structure formed from the maturation of an ovule and containing an embryo and stored food, pp. 13, 26, 84, 89, 160, 164.

Seed coat. The protective outer layer of a seed, p. 26.

Seed dispersal. pp. 26, 87, 180.

Seed germination. See germination.

Seed leaf. See cotyledon.

Seedling. A young plant, shortly after seed germination, p. 29.

Self-pollination. The transfer of pollen from an anther to the stigma of the same flower, p. 173.

Senescence. The aging process; a breakdown of cellular structures leading to death, pp. 83, 127.

Sepal. A flower part that usually encloses and protects the flower bud, p. 162.

Sessile leaf. A leaf in which the blade is directly attached to the stem, p. 45.

Shade-tolerant. Having the ability to live in low light intensities, pp. 100, 123.

Shoot. A stem bearing leaves, pp. 16, 39.

Short-day plant. A plant flowering in response to days shorter than its critical photoperiod, p. 134.

Shrub. A woody plant with little or no trunk and having branches near its base, p. 42.

Sieve plate. The perforated end-wall of a sieve tube member, p. 72.

Sieve tube. A food-conducting cell, p. 72.

Silt. An inorganic soil component the particles of which range between 0.002 and 0.02 mm in diameter, p. 149.

Simple fruit. A fruit formed from one ovary, p. 179.

Simple leaf. A leaf in which the blade is not divided into smaller units (leaflets), p. 47.

Sorus. An area of spore production on the underside of a fern leaf, p. 189.

Spadix. A spike of flowers enclosed in a spathe, (photograph) p. 171.

Spathe. A large bract enclosing a spadix, (photograph) p. 171.

Species. A group of individuals sharing many characteristics and interbreeding freely, pp. 77, 182.

Specific epithet. A taxonomic classification; the second part of a species' scientific, binomial name, p. 183.

Sperm. A male sex cell, pp. 160, 176, 187, 189, 191.

Spike. An inflorescence in which the flowers are attached to the main stem without stalks, (diagram) p. 169.

Spine. A modified leaf part that is hard and sharply pointed, pp. 87, 88.

Spongy cell. One of a group of loosely-packed photosynthetic cells in a leaf, p. 69.

Spore. A reproductive cell that grows directly into a new plant, pp. 13, 160, 189, 191.

Sporophyte. A diploid, spore-producing plant in an alternation of generations, pp. 189, 191.

Springwood. Xylem laid down by the vascular cambium in spring and early summer, p. 61.

Spur. A tubular projection from a flower, p. 167.

Stamen. The male part of a flower, consisting of an anther and filament, p. 164.

Starch. The principal food-storage substance of higher plants; a carbohydrate consisting of numerous glucose units, p. 155.

Stem. The leaf- and flower-bearing part of a plant, pp. 39, 43, 54, 62, 101, 103.

Stem tuber. An enlarged tip of a rhizome containing stored food, p. 110.

Stigma. The part of a pistil that receives pollen, p. 164.

Stilt root. See prop root.

Stinging hair. A multicellular hair containing an irritant fluid, p. 89.

Stipule. An outgrowth from the base of a leaf stalk; sometimes functioning as a protective structure, (diagram) pp. 46, 88, 103.

Stock. A rooted plant into which a scion is inserted during grafting, p. 60.

Stolon. See runner.

Stoma (stomata, plural). A pore in the epidermis of leaves and herbaceous stems, pp. 69, 142, 143.

Stone cell. A hard, thick-walled plant cell, p. 74.

Stratification (to stratify). A cold treatment given the seed of some species to improve the percentage of germination, p. 33.

Style. The narrow part of a pistil bearing the stigma, p. 164.

Suberin. A fatty plant substance present in the walls of cork cells, pp. 59, 84, 90.

Sucker. An adventitious shoot arising from a root, p. 101.

Summerwood. Xylem laid down by the vascular cambium in late summer, p. 61.

Symbiosis. The living together for mutual benefit of two or more organisms of different species, pp. 90, 113.

Tannin. A substance occurring in the bark or leaves of some species, functioning to protect against predators, pp. 90, 93.

Tap root. A prominent root with few branches, sometimes swollen to store food, p. 35.

Tendril. A modified stem or leaf for climbing, pp. 103, 125.

Tepal. A perianth part in flowers having no distinct petals or sepals, p. 164.

Terminal bud. See apical bud.

Terminal bud scale scar. A scar left on a stem after the bud scales have fallen, p. 44.

Thigmotropism. A growth response to touch, p. 125.

Thorn. A modified stem that is hard and sharply pointed, p. 87.

Tissue. A group of cells of the same type having a common function, pp. 52, 53.

Totipotency. The capacity of certain cells, when isolated and properly grown, to regenerate a whole plant, p. 72.

Trace element. See micronutrient.

Tracheid. A water-conducting cell in Gymnosperms and other lower vascular plants, p. 73.

Transpiration. The loss of water vapor from a plant, mostly from the stomata of leaves, p. 142.

Transpirational pull. The force exerted by transpiration from the leaves that draws water up through a plant, p. 142.

Tree. A large, woody, perennial plant having a definite trunk, p. 42.

Tropism. A growth curvature of a plant part caused by some external stimulus such as light or gravity, p. 120.

Tuber. See root tuber; stem tuber.

Tuberous root. See root tuber.

Turgid. Swollen and firm due to internal water pressure, p. 137.

Turgor pressure. The pressure developed in a cell as it becomes filled with water, p. 139.

Twiner. A stem growing in a spiral fashion around a supportive object, p. 103, 126.

Umbel. An inflorescence in which the flower stalks arise from one point at the tip of a stem, (diagram) p. 170.

Vacuole. A fluid-filled sac within a cell, pp. 20, 137.

Variegation (variegated). An inherited, irregular pattern of color in a leaf or petal, p. 151.

Vascular bundle. A strand of conducting tissue containing xylem and phloem, pp. 55, 62.

Vascular cambium. A narrow cylinder of cells that gives rise to secondary xylem and phloem; a lateral meristem, pp. 56, 57, 66, 67, 84.

Vascular plant. Any plant containing water- and food- conducting tissues, p. 75.

Vascular ray. A narrow sheet of cells running radially across the secondary vascular tissues of a stem or root, p. 60.

Vascular tissue. A group of food- or water-conducting cells, pp. 55, 66.

Vein. A strand of xylem and phloem in a leaf blade, pp. 48, 69.

Velamen. A water-absorbing tissue on the outside of orchids' aerial roots, p. 106.